ERWIN SCHRÖDINGER'S WORLD VIEW

THEORY AND DECISION LIBRARY

General Editors: W. Leinfellner (*Vienna*) and G. Eberlein (*Munich*)

Series A: Philosophy and Methodology of the Social Sciences

Series B: Mathematical and Statistical Methods

Series C: Game Theory, Mathematical Programming and Operations Research

Series D: System Theory, Knowledge Engineering and Problem Solving

SERIES A: PHILOSOPHY AND METHODOLOGY OF THE SOCIAL SCIENCES

VOLUME 16

Scope: This series deals with the foundations, the general methodology and the criteria, goals and purpose of the social sciences. The emphasis in the new Series A will be on well-argued, thoroughly analytical rather than advanced mathematical treatments. In this context, particular attention will be paid to game and decision theory and general philosophical topics from mathematics, psychology and economics, such as game theory, voting and welfare theory, with applications to political science, sociology, law and ethics.

The titles published in this series are listed at the end of this volume.

ERWIN SCHRÖDINGER'S WORLD VIEW

The Dynamics of Knowledge and Reality

edited by

JOHANN GÖTSCHL
Department of Philosophy,
University of Graz, Austria

KLUWER ACADEMIC PUBLISHERS
DORDRECHT / BOSTON / LONDON

Library of Congress Cataloging-in-Publication Data

Erwin Schrödinger's world view : the dynamics of knowledge and reality / edited by Johann Götschl.
p. cm. -- (Theory and decision library. Series A, Philosophy and methodology of the social sciences ; v. 16)
Includes bibliographical references and indexes.
ISBN 0-7923-1694-0 (alk. paper)
1. Schrödinger, Erwin, 1887-1961. 2. Physics--Philosophy. 3. Life. 4. Self-organizing systems. 5. Philosophy of nature. I. Götschl, Johann. II. Series.
QC16.S265E76 1992
500--dc20 92-7367

ISBN 0-7923-1694-0

Published by Kluwer Academic Publishers,
P.O. Box 17, 3300 AA Dordrecht, The Netherlands.

Kluwer Academic Publishers incorporates
the publishing programmes of
D. Reidel, Martinus Nijhoff, Dr W. Junk and MTP Press.

Sold and distributed in the U.S.A. and Canada
by Kluwer Academic Publishers,
101 Philip Drive, Norwell, MA 02061, U.S.A.

In all other countries, sold and distributed
by Kluwer Academic Publishers Group,
P.O. Box 322, 3300 AH Dordrecht, The Netherlands.

Printed on acid-free paper

02-1292-250 ts

Printed in the Netherlands

ACKNOWLEDGEMENTS

The editor has not only the obligation but also the pleasure to thank everyone who contributed to this volume. We, at the Ludwig Boltzmann Institute for Science of Research, began work on this book project in 1987.

I would like to express my deep appreciation to Werner Leinfellner. He contributed substantially to giving this book its conceptual orientation and structure. Wolfgang Schinagl, our research assistent, created the camera-ready type version. Günther A. Höfler read and corrected the texts. Angelika Höber, the Institute secretary, provided invaluable services in preparing the material for print. Camilla R. Nielsen made editorial suggestions and reworked the texts for the English version of this book. My thanks go to those named here and to the authors and publisher for all their cooperation in helping us to complete this book.

Furthermore I should like to thank Auguste Dick, Gabriele Kerber, Wolfgang Kerber and Karl von Meyenn, who have edited and completed the bibliography with utmost care. Thanks to their efforts, it has also become an excellent book of reference.

Erwin Schrödinger during a lecture in Graz around 1937.
Photo: Heinz Reuter

TABLE OF CONTENTS

PART III

ERWIN SCHRÖDINGER'S WORLD VIEW AND HIS PHILOSOPHICAL SYSTEM

INTRODUCTION

JOHANN GÖTSCHL
Ludwig Boltzmann Institut für Wissenschaftsforschung und Institut für Philosophie der Universität Graz Mozartgasse 14, 8010 Graz, Austria

WERNER LEINFELLNER
Department of Philosophy University of Nebraska, Lincoln, NE 68588, USA

Erwin Schrödinger is well-known as one of the greatest figures of theoretical physics. He became famous for his mathematical formulation of atomic dynamics, known as the Schrödinger-equation, and its application to atomic structure. In 1933, he received the Nobel Prize in physics (jointly with Paul Dirac). It therefore comes as no surprise that interpreters of Schrödinger's work, dazzled by his genius and the results he obtained in quantum physics, almost forget the other Schrödinger, the biologist and philosopher. In reality, he not only revolutionized the foundations of modern physics but also made significant contributions to biology and the philosophy of science. The paradigm change he initiated has had great consequences for the natural and social sciences, the humanities, and our present scientific world-view in general.

It is not so well-known that there are such powerful principles behind Schrödinger's wave mechanics, such as his dynamic view of matter and his holistic view of nature. Such views mark the beginning of our era of dynamic sciences and our new dynamic world-view with it ongoing synthesis of quantum physics and quantum chemistry. In his book *What is Life?*, Schrödinger tried to integrate quantum physics and quantum chemistry with genetics and evolutionary biology. His dynamic view of the evolution of life has led to the present breakthrough in the biology of evolution, to the greatest intellectual adventure in the history of science and culture: the deciphering of the genetic code, followed in the next decade by the deciphering of the human genome. The latter development will give rise to a new, scientific meaning of Socrates' "Know thyself". The dynamic concept of evolution has had a strong bearing on today's unified theories of the universe, such as the GUT theories. With these theories we are able, for the first time in history, to explain the origin and evolution of our universe and the beginning of self-organization in the life of civilizations and cultures. The final theorization of these dynamic evolutionary processes would not have been possible without Schrödinger's pioneering work.

The present book seeks to bridge the gap between Schrödinger the physicist and Schrödinger the philosopher and to show how his original and provocative views - formulated in the precise language of a physicist - contributed to revolutionary changes in the philosophy of science and in science as a whole. One consequence of this new outlook is the present integration of natural sciences, biology, sociology, and even religion and culture in a new science of life and of evolving societies.

The articles in this book will show, in terms understandable even to the non-physicist, the fundamental changes that have shaken the traditional sciences and our traditional view of the

J. Götschl (ed.), Erwin Schrödinger's World View, 1–14.

world. It will become clear how these changes originated from Schrödinger's pioneering thoughts and in what ways they affected both the foundations of both physics and modern cognitive science. In other words, this book describes a status nascendi, the birth of the new dynamic orientation of modern science which reflects the spirit of a new physics and its philosophy. Contrary to Kuhn's view, the introduction - and the book as well - do not theorize abstractly about paradigm changes. The book deals with an ongoing paradigm change which has shaken not only the foundations of physics, but also, through a sort of snow-ball effect, most of traditional cognitive science.

As soon as Schrödinger's wave mechanics appeared in print, a pandora box of problems and paradoxes was opened, one of which was the famous paradox of Schrödinger's cat. And this is something the reader will soon notice. This is part of the intellectual heritage which Schrödinger bequeathed to physics and modern science as a whole; it did not die and perhaps never will. It remains a challenge to us since it marks the transition from old and established concepts to new ones. This paradox always surfaces, when new thoughts about reality clash with the views of a scientific community which did not have the courage to debunk the old, traditional views.

This is exactly what happened to physics as soon as Schrödinger's wave mechanics, a new dynamic model of matter, appeared. It unleashed a world-wide dispute regarding its foundations and consequences. It brought up again, for example, the old nagging question about the reality of our external world. Could the familiar, millennia-old Aristotelian concept of an underlying, unchangeable substance, e.g, the atom, be abandoned in favor of Schrödinger's new dynamic one where "substance" resembles a bundle of waves or melodies? This, at first, seemed impossible. Indeed, Schrödinger's turn to a universal dynamics can be seen as unsettling: change is now the main characteristic of reality. That is to say, nature and society are actually in a constant state of flux and disturbance. This makes it very difficult to explain how order and stability came into existence and and can coexist with randomness and chaos. Precisely this is the crux of Schrödinger's philosophy.

The book will trace the development of thought from the dynamics of microparticles and atoms to the dynamics of living macromolecules. His views, in turn, have led his adherents to the evolutionary dynamics of life in our cosmos, to the self-organization of order. They also have had consequences for the evolution of societies, technologies, sciences, and cultures.

This book has two main strands: (i) One part of the book deals with Schrödinger's personal philosophy and his fight against traditional ideas; they describe how and why these ideas motivated him to proceed from mathematics and quantum mechanics to thermodynamics, to biology, and, finally, to philosophy. (ii) Other articles analyse the impact of Schrödinger's turn and of his new ideas on all sciences and our world view. His ideas, in fact, continue to influence our present theories about the evolution of the cosmos, of life on earth, of consciousness, and culture. They underline the responsibility of man's responsibility vis-à-vis science, technology and even evolution.

To understand Schrödinger's controversial, even polemical ideas it is necessary to grasp why they had such an impact on his contemporaries, on today's philosophy of science, and on our modern world-view. His ideas were, as he said in his address to the Prussian Academy of Science in 1922, "a result obtained from a lifelong love of science." During his last two years in Vienna, a local philosopher (who contributed to this volume) asked him what he meant by "love of science" and whether he regarded himself as a physicist or as a philosopher. Schrödinger responded, in his humorous way (which Halpern, his former assistant, describes so well in his article): "you know that Pythagoras is said to have called himself a lover of wisdom." Anyone who has ever read Schrödinger's remarkable book on Greek philosophy will understand what he meant by this.

Schrödinger's thinking unites the holistic view of wave mechanics and the dynamics of living beings, individuals and culture and anticipated developments in practically all of modern science. In brief: we first have to understand the world, the sciences, and culture as a dynamic and interdependent whole. This basic insight then led Schrödinger, and anyone who reads him closely, to new views of central problems of philosophy and life, reality, the mind-body problem, the nature of the 'I' and of consciousness, and to a new answer to the question of the meaning of life. But if we really want to follow Schrödinger, we have to be willing to free ourselves, just as Schrödinger did, from the fetters of our traditional, exclusively discursive European way of thinking. The East seems to provide a more intuitive and holistic approach, one which could ultimately be adopted. Schrödinger liked to compare Western and Eastern philosophy with the idea of split brains. The left side of our brain perceives the world in a linear, logical, and rational way, and can be seen as corresponding to the yen of Chinese philosophy. The right hemisphere perceives patterns as a whole, it works intuitively; and it corresponds to the Chinese yang. It seems that Schrödinger's philosophical convictions and his new holism can be illustrated by the following idea: in spite of the fact that both hemispheres work in a different way, they cooperate when thinking and solving problems.

To illustrate the unity of body, mind, and consciousness Schrödinger borrowed holistic ideas from the East. But to his own amazement, he discovered that his new holistic concepts clashed not only with Western thought but also with the fundamental laws of traditional Western science.

Nolens volens, Schrödinger became the father of all those scientists and philosophers who have tried to integrate Eastern holistic solutions into the "Lord's quantum mechanics" - as he himself said in 1948. But he also suggested that this Eastern holism could be adopted for solving global problems, e.g., world-wide pollution, deterioration of our planet's climate, nuclear energy, etc. Schrödinger's ideas are also relevant for the unification of East and West which will unite a multitude of cultures in a future world culture. It becomes more and more evident that we cannot simply rely on the competitive way of thinking of the West.

The underlying philosophy of the articles collected in this volume is consistent with the views held by a growing number of physicists, psychologists, biologists, social scientists, and philosophers. If we read Schrödinger closely, we will see the confirmation of a truth stated earlier in this introduction: The greatest hindrance to understanding Schrödinger is not his new ideas, but our reluctance, or even inability to give up received views and traditional concepts. It is thus no wonder, then, that Schrödinger's revolutionary views and his criticism of traditional thoughts were suppressed and persecuted by the Nazis. (Schrödinger left Berlin in 1933 shortly after Hitler came to power and Graz in 1938 when Austria became part of the Third Reich). As a result of German politics Schrödinger's philosophical works disappeared from public consciousness in the German-speaking countries. The immense progress of physics, biology, and social science today shows that his revolutionary philosophical ideas were topical. Schrödinger can be regarded as the spiritus rector of the present turn in our modern scientific world view; his ideas have triggered a revolution in today's philosophy of nature and philosophy of mind, in biology, psychology, and recent ideas in the social sciences.

In *Erwin Schrödinger - a Biographical Sketch*, an appendix to the introduction, Johann Götschl and Günther Höfler provide a brief overview of Schrödinger's life as a scientist and researcher. Following Schrödinger, who strongly disliked chronologically ordered biographies, they focus on the most important individual experiences and periods which had a lasting impact on his life.

SURVEY

The 14 papers which have been selected for this volume attempt a new philosophical look at Schrödinger's work. They also deal with the impact of Schrödinger's work on today's cognitive sciences, on philosophy, and on our world view.

The papers are divided into three major sections:

1. Schrödinger's work as a physicist and its philosophical impact
2. Schrödinger's theory of life and self-organization
3. Schrödinger's significance for contemporary philosophy, i.e., natural philosophy, social philosophy, and our present-day world view.

The beginning of this century ushered in an unprecedented era of turnover and re-evaluation of all classical sciences that had governed the nineteenth century. The first group of papers in the first section addresses this new relationship between quantum physics and philosophy of nature and its impact on philosophy of science. These papers also elucidate why the turn from classical physics to quantum physics had such a tremendous impact on the change from traditional static sciences to dynamic ones. It seems that today's cognitive sciences are beset by the same or very similar problems which continue to plague the foundation of quantum physics. Our only advantage is that we can learn from the history of quantum theory that it has provided us with means to solve them.

In *Erwin Schrödinger and the Philosophy of the Physicists,* E. Scheibe examines all the intricate epistemological problems of Schrödinger's wave dynamics, as for example the revolutionary view of matter as time-dependent and dynamic in Schrödinger's equation. Quantum mechanics was the first genuine "theory of change". Its assumptions about dynamic structures are of prime importance for our century because modern cognitive sciences appropriated this "dynamization". Schrödinger's wave mechanics broke with the traditional, orthodox static view of reality. Scheibe illustrates this by pointing to the inherent conflict between Schrödinger's objectivation postulate (his reality postulate), Schrödinger's own wave dynamics and our commonsense understanding of nature and causality. Schrödinger saw very clearly that his new foundations of physics would sooner or later clash with our traditional views of the world and lead to a philosophical gigantomachy vis-a-vis the reality of our world. In a nutshell: Is nature in a state of constant flux or is there such a thing as never-changing objects? This conflict could not be solved within the field and means of physics and even went against the grain of all traditional physical laws. In particular, Schrödinger's concept of inseparability (holism), his concept of objectification (reality postulate), his substitution of deterministic laws through statistical laws are still topical today - ideas which force us to abandon our classical lines of thinking. These changes induce new epistemological and ontological paradoxes, uncanny follow-ups of the cat paradox.

In *Holistic Aspects of the Schrödinger Equation*, H. Pietschmann discusses several rifts which run through Schrödinger's work. For instance, when he defended his own wave mechanics and its statistical interpretation, Schrödinger opted for neither the pros nor the cons of his theory. Being a meticulous and modest scientist, he turned to contemporary physicists for help. And there is yet another rift: the mathematical, deterministic nature of Schrödinger's wave equation as opposed to its indeterministic, statistical interpretation. Schrödinger also liked to think "metatheoretically" about theories in terms of classical "either-or" categories. Pietschmann sees this as a reflection of Schrödinger's personality. This either-or approach made it impossible for Schrödinger to make a clear choice between the alternatives: i.e, either to accept all the consequences of his own wave mechanics, including its new and strange holistic world view, or to stick to Einstein's classical, traditional world view. These discrepancies resulted in alternatives which continue to plague quantum theorists today. The reader can opt for either of the following:

1) We can picture and represent our reality, as in the classical sciences, which includes, of course, Newtonian physics or we can describe reality only statistically and mechanically and cannot represent it in a "pictorial" way, as in quantum physics.

2) We can describe things and their changes precisely in a local space and time, as in the classical sciences or we can observe the behavior of subatomic particles only statistically (indirectly and inprecisely), as in quantum physics. The latter can be applied to the behavior of individuals in a society as well.

3) We can predict individual events, as in the classical sciences, or we can predict frequencies, probabilities in group behavior, as in quantum physics. The latter option can be applied to the social sciences.

4) We can assume an objective reality "out there", as in classical physics and the classical sciences, or we can say that there is no objective reality that is totally detached from our experiments and observations, as in quantum physics.

5) We can observe and measure something without changing it, as in the classical sciences and in everyday life, or we cannot observe anything without changing it, as in quantum physics.

6) We can agree with the classical sciences that "absolute truth" is possible and that, in keeping with Einsteins's view, physics describes nature as it really is, or, we can say that there are only dynamic relations which we express in terms of statistical correlations between events, individuals, etc. "behind the scenes," as in quantum physics.

When confronted with these alternatives, Schrödinger chose to follow neither Boltzmann, Einstein, Rosen, Podolsky, etc., nor any of the "Copenhagen" physicists Bohr, Born, Heisenberg, etc. In this greatest scientific debate about reality and the foundation of our knowledge of the world, he positioned himself between the two antagonistic groups.

Pietschmann, however, shows convincingly, and elaborates in mathematical details, that Schrödinger actually went beyond Einstein by introducing this new holistic and dynamic aspect of quantum mechanics. Pauli, Bell, and Aspagnat, and later the stochastic evolutionists later took up, and confirmed, his solution. Thus Schrödinger's holism turns out to be an underpinning of modern philosophy. This holistic view explains his new concept of reality better than anything else. It influenced his interpretation of quantum mechanics and the development of quantum field and gauge theories.

Anyone who thinks that it is no longer possible to write an article about Schrödinger's celebrated "cat" after the flood of articles that have appeared since 1935 is mistaken: In *Schrödinger's Cat and the Interpretation of Quantum Mechanics*, B. Kanitscheider interprets Schrödinger's famous cat "experiment" as a metaphor of Schrödinger's and our basic philosophical views of reality and science.

Schrödinger's thought experiment of 1935 demonstrated the paradoxical consequences of quantum-physical indeterminacy for macroscopic reality. Schrödinger was a cat lover; and

Schrödinger's experimental cat was, of course, an imaginary cat in an imaginary experiment. Nevertheless, this cat almost shattered quantum physics, as Kanitscheider states. Even today, this imaginary experiment challenges us to decide between its two possible solutions concerning the cat's reality after the animal has been locked up in a container for an hour. Schrödinger illustrated the role of chance for all dynamic and evolutionary processes as follows: the cat's existence depends on a random event (radioactive disintegration) which has a 50% chance of occurring (higher than in atomic incidences). After an hour this results in the following paradox: the cat is in the ghostly, paradoxical situation of being 50% alive and 50% dead - a state equivalent to the indeterminate one of microparticles in their atomic environment. But the cat can only be either 100% dead or 100% alive.

With the help of this "experiment", Kanitscheider describes the most famous controversy of the twentieth century, the one between the Copenhagen school on the one hand, and Einstein, Rosen and Podolsky on the other - and Schrödinger in between. It is well-known that the controversy revolved around the interpretation of quantum theory and the new holistic features of reality which tie the observer to the observed system. Schrödinger himself was the first to give a paradoxical description of a macrophysical body solely in terms of microphysical properties, i.e., by the ψ function. He felt uneasy about the consequences of his dynamic theory and its hidden holism: a subjective quality or activity of human beings, for example, observations, measurements, the semantics of their languages, or their consciousness can influence nature's physical state.

Kanitscheider elegantly uses Schrödinger's "cat" as a paradoxical paradigm for the new holism contained in the superposition principle of quantum mechanics in order to illustrate all the problems which have haunted science from Schrödinger to modern cosmology. For example: Is it only paradoxical if our common sense describes the state of the cat after an hour in terms of a wave function, i.e., by the superposition of two substates, dead and alive? Does this make any sense for macroscopic bodies and for our physiological, everyday perception of reality? Or does our reality have features which are alien to our daily experience? But even if we accept the new alien features of reality, we still encounter almost insurmountable difficulties with today's quantum-physical interpretations, e.g., Wigner's or Wheeler's many-worlds interpretation, and Barrow's and Tipler's anthropic principle. Kanitscheider deals with these in detail.

For Kanitscheider, the crucial question is: should we regard the transition from superposition to single states as an idealistic epistemic process (in the mind of the observer, i.e., our mind) or as an ontic, empirical step (process) in physical reality?

Kanitscheider sees today's evolutionary physics as confirming Schrödinger's original philosophical realism. He expects the remaining riddles to be solved by introducing gravitational interaction in quantum mechanics as was recently proposed by Penrose. This could restore objectivity and confirm Schrödinger's realism of the external world.

The last paper in this section, K. Bauman's *Schrödinger's Function of Macroscopic Systems*, continues the discussion of the "cat or reality problem" which is so important for the question as to what extent our macroscopic world is influenced by quantum physics and its laws. Some of them, no doubt, are influential, as for instance, mutations. Recent research has proven that, under certain well-defined conditions, macrophysical bodies still lacking hierarchical structure, as for instance, living organisms, react, in fact, like microphysical ones. This question has been addressed again in various new theories on macroscopic quantum coherence; they operate with SQUIDs (supra-conducting quantum-interference devices), jokingly called "artificial Schrödinger cats" by Flamm. This issue is also dealt with in theories about superconductivity and the tunnel effect. Baumann shows that these newest results confirm the validity of Schrödinger's equation, but only for specifically structured macroscopic bodies. His interpretation facilitates a better understanding of Schrödinger's position between the Einsteinian and the Copenhagen interpretation of reality.

The second section sums up results and controversial issues of Schrödinger's quantum mechanics, exploring its consequences for the biological and the social sciences, and our modern concept of life. The articles show how the dynamic concept of evolving life, which dominates our modern view of life and culture, originated from Schrödinger's work in physics, biology, and philosophy.

Rupert Riedl (*Schrödinger's Negentropy Concept and Biology*) is fascinated by the fact that Schrödinger was both a physicist and a biologist. He is convinced that only future cooperation between physicists and biologists can help us to successfully continue Schrödinger's quest for a final answer to the question "What is life?" Riedl contends that Schrödinger overemphasized the role of entropy and order, while neglecting the concept of free energy and energy exchange between evolving systems and their environment. It is the latter concept which is becoming more and more important for all life sciences and life on earth as well.

Riedl begins with Schrödinger's dynamic equations: increase of negentropy = increase of order; increase of entropy = increase of disorder. Riedl points to the precursors of Schrödinger's famous book "What is Life?" (e.g., Boltzmann) and shows what has followed it in recent years. If it is the upshot of quantum physics and entropy that our cosmos is not only in a constant dynamic change but also increases disorder, how can we explain the course of the evolution of life and the relative stability of the species in a quantum-physical world? Or, to put it differently: How can a biologist explain the dynamics of evolution, the self-organization of life and the relative stability of all living species? Riedl shows convincingly that, if biological order is understood not only as dynamic but also as evolutionary, it is much easier to explain it in terms of equilibrating energy-conservation and energy-waste between living systems and their environments than by the concept of order. This energy concept has been formulated recently by Prigogine. Together with the function of hierarchical structures of organisms and the role of information for living organisms, it complements Schrödinger's view. For the first time, we are now able to explain the evolution of the world from primordial chaos in terms of the increase in storage capacity for genetic information and instructions concerning the self-organization of higher forms of life and ultimately of social systems.

At the end of his article, Riedl draws some conclusions for our civilization as endangered species. If we fail to learn from these ideas, as for example how to equilibrate energy exchanges between humans, other terrestrial living systems, and their environments, we will inflict economic and socio-cultural destruction, environmental catastrophes, etc. on our planet.

In *Determination and Self-Organization - Erwin Schrödinger's Views on Chance*, H. Hörz focuses on the eminent role and function of probability and chance in Schrödinger's quantum mechanics and in modern cognitive sciences. He shows that Schrödinger's simple model of the self-organization of relatively stable open systems - including societies - cannot stand the test without chance and randomness. To be sure, Schrödinger adopted the concept of chance in the sense of an active force from his predecessors, e.g., Laplace. He was, however, the first thinker to put chance in the wider context of dynamic, evolutionary processes.

This again led to new problems in understanding dynamic processes. Are they self-organizing, non-teleological, evolutionary processes (Prigogine, Eigen, Haken, Dawkins, Wuketits), or are they deterministic, teleological ones (Einstein, Rosen, Podolsky)? According to Hörz, it all depends on the role one sees chance as playing. On the one hand, self-organization, the crucial factor in all dynamic, evolutionary processes, cannot work without chance; this inevitably means that causal laws are less powerful. Otherwise there would be no innovations, no new ideas, and no new species. On the other hand, strictly deterministic processes and their traditional formulation as mechanistic, causal laws exclude chance.

Schrödinger opted for chance, thereby initiating a completely new stochastic theory of evolution (the theory of dynamic differential games, as proposed by Eigen, Schuster, Maynard-Smith, Leinfellner, etc.). This is the only theory that explains evolutionary processes, using a new type of statistical, stochastic laws which were introduced by Schrödinger. Moreover, this theory is equally well-suited for explaining natural and social evolutionary dynamic processes. Here the contradiction discussed above is resolved.

At the end of his article, Hörz proposes six philosophical principles which, according to him, play a leading role: self-organization, the social principle of cooperation, hierarchical order, the pursuit of relative goals, socio-ethical control, and regulation of evolutionary processes by man.

In *Schrödinger, the I and the Genes*, W. Leinfellner - who happened to know Schrödinger personally between 1956-1960 - presents Schrödinger as the first scientist to fully appreciate the scientific, biological, and socio-cultural significance of the genes and the new, scientific meaning which they gave to Socrates' "know thyself." For Schrödinger, the genome turned out to be an order-preserving, negentropic empirical factor; it represents, maintains and regulates the highest form of order on earth. At the same time, genes are a monumental codified, intersubjective memory of life's past on earth; they determine, albeit in a statistical sense, present and future evolution. Genes guarantee epigenesis, growth, metabolism, replication, and the relative stability and self-identity of living organisms throughout their lifetime. Genes are also the codified seat of Schrödinger's higher laws; they are superimposed on physical laws and cannot be reduced to the ordinary laws of physics.

As we have already seen, genes regulate not only the metabolism, the construction of our body and our brain, replication, and our basic social behavior; they also maintain the self-identity and stability of the ego, or what our Western philosophy and religion calls "the soul". Thus the self-identity and stability of individuals is encoded in, and guaranteed by, our genes; but self-identity and stability are relative since they last only as long as the individuals live. For Leinfellner, Schrödinger's concept of negentropy and Prigogine's nonequilibrium theory anticipated the evolutionary stability of organisms.

Schrödinger's introduction of the concept of an order-preserving genome into biological evolution leads right up to Maynard-Smiths' stochastic theory of evolution, the former being the missing link between the modern theory of evolution and Darwin's theory. It is no wonder, then, that Schrödinger foresaw as early as 1944 that quantum physics and chemistry would merge with biology and genetics. The result was to be a comprehensive new scientific theory of evolution which would change our view of man and nature.

In the last chapter of his article, Leinfellner outlines Schrödinger's philosophical views on the interrelations between the self, the genes, and our personality (c.f., Poser's contribution). Since the stability of our genes which guarantees the stability of our personality is only relative, that is, ends when our lives end, Schrödinger was drawn into a controversy over our Western and

Christian ideologies which assume the existence of an immortal, immaterial soul. It is again no wonder that Schrödinger - as many physicists after him - turned to a form of Eastern (Indian) philosophy which equates the personality, the ego, or the soul with the self-identity of the individual during his/her lifetime. Consequently, for Schrödinger consciousness or the ego is only a singular experience of this identity forced upon us by the memory of our past life. For him - and for Freud - neither a plural nor a concept, the ego (as it is usually called), exists. The ego exists only in the singular, as my unique ego.

In his article, *Erwin Schrödinger and his Contribution to an Understanding of Living Systems*, F. W. Wuketits describes how Schrödinger found the principles of a new synergetic approach to the science of life and society in the holism of quantum theory and in its dynamics to integrate the hitherto unconnected natural sciences of the nineteenth century into a new, hierarchically structured, synergetic whole, a new science of life and of society. In the twentieth century, this new approach developed into the theories of cosmic evolution. Schrödinger's "dynamization" was the spark that triggered the gradual integration of quantum chemistry into quantum physics. These, in turn, were integrated into a dynamic biology of evolution. Schrödinger developed his ideas without succumbing to the risk of reducing life to chemical or physical processes. According to Wuketits, this was one of his greatest achievements.

Thus, Schrödinger's ideas have helped form the cornerstone of a new science of the evolution of life, man, society and culture. This new science offers solutions for problems such as: How does living matter evade decay or entropy? What is the past and present role of hereditary substance and the genes? What is the nature of consciousness? Why is our will free? Should mankind take socio-cultural progress into its own hands? According to Wuketits, future evolution can only be steered by mankind itself with its new insights into the course and the nature of biological evolution.

Wuketits addresses many issues also discussed by Riedl, Leinfellner, and Hörz. His focus, however, is somewhat different and more general. His intention is to show what we have learned from our knowledge of biological evolution. Wuketits, one of the founders of a new epistemology, the evolutionary theory of knowledge, sees the development of Schrödinger's ideas up to the present as an example of the evolution of knowledge. This ongoing process of evolution entered its decisive phase with Schrödinger. For him, Schrödinger was the great synthesizer who built the basis and the theoretical framework, not only for molecular biology and a modern theory of evolution, but also for our biological, philosophical and cultural self-understanding.

The third section, "Schrödinger's Importance for Contemporary Philosophy, Philosophy of Nature, and Our World View", centers on an original epistemological idea of Schrödinger.

Most interpreters see Schrödinger as the architect of our modern "external", scientific world-view. This is simply because the objective bent of his scientific view of our world has been discussed widely and is accessible to all of us through books and libraries. But this is only one side of his epistemology. Schrödinger, the scientist, whose profession was to supervise the construction of the new foundation of our objective, external view of nature, quantum theory, became more "introspective" in his later life.

Whereas the first two chapters as well as Halpern's article deal primarily with the objective bent of Schrödinger's scientific world-view, the articles by Elkana, Röseberg and Poser, but,in part, also Leinfellner's article, discuss Schrödinger's introspective epistemology - an optic which throws light on individual and personal cognition and on the acquisition of our individual world view. The third section focuses on the intimate question of how the common world view, constructed by the community, the teamwork of scientists, philosophers, artists, and all other concerned persons, becomes my own world view.

Leopold Halpern, a former assistant of Schrödinger, discusses his personal experience of Schrödinger and Schrödinger's impact on the development from physics to today's modern cosmology in *On the Scientific and Humanistic Thinking of Erwin Schrödinger During his Last Years*. During Schrödinger's last years in Vienna, Halpern had the invaluable privilege to be in personal contact with him and to discuss many problems with him. The paper throws a new light on Schrödinger's personal comments, Schrödinger's criticism of his own theory, and on the development of modern cosmology (Hawking, Penrose). Most characteristic for Schrödinger the scholar was his tolerant, but at the same time critical philosophical attitude, even towards his own quantum mechanics. Halpern sees the paradigm change, as described in the introduction, as beginning at the moment Schrödinger met his teacher Exner. When Schrödinger cast Exner's dynamic views into the mathematical form of the famous Schrödinger equation, he became immediately aware of the change he had brought about in the conceptual and formal foundations of classical physics. Schrödinger was convinced that this change would sooner or later spread to the very fundamentals and methods of all cognitive sciences, in the same way relativity theory had changed our classical views about space and time and traditional cosmology. Halpern, calling up his personal memories of Schrödinger, clearly sees the inner ties of Schrödinger's dynamic world view and his holism with the "cosmological revolution" of today's modern physics, for example, Hawking's and Penrose's theory of black holes. It is interesting that, according to the author, Schrödinger was well aware of the theoretical possibilities of black holes and neutron stars; but he remained skeptical about the validity of any possible consequences which could not be empirically tested. For Schrödinger, science should always be open to empirical tests. As long as its results cannot be confirmed or rejected, it is not a cognitive science. He could not, of course, foresee the later tests of Hawking's theory of black holes.

In *Erwin Schrödinger as Historian - Notes Towards an Interpretation*, Yehuda Elkana examines Schrödinger's work in its historical context and from the vantage point of Schrödinger's search for unity in the world and in science - which is just another version of his dynamic holism. For Elkana, Schrödinger was at the same time a great scientist, philosopher, humanist, and historian. As historian he not only wrote a book called *History of Greek Philosophy* but also always analyzed the results and views of his predecessors in their historical contexts. Schrödinger also saw a close relationship between physics and history since both deal with indeterministic and irreversible processes which cannot be explained by deterministic causal laws. This is quite contrary to Einstein's view.

When Elkana applied Schrödinger's relativistic historical view to Schrödinger himself, he discovered that, on the one hand, Schrödinger was deeply influenced by certain rational, empirical views which he found in Vienna's academic circles (Exner and other scientists). On the other hand, his own views may be linked to those of the "uprooted" poetic and intuitive Austrian cosmopolites and the "fin-de-siecle" prophets like Kafka, Kraus, among others. Schrödinger always tried to bridge the gap between these two Viennese cultures. Their marriage was to give rise to a new world view, a new humanism beyond rationalism and traditional humanism. For example, since both "cultures", as well as all empirical sciences, are governed by statistical laws this would guarantee all individuals a relative leeway and freedom within democratic societies.

Schrödinger saw very clearly that free will and ethics cannot be subjected to either deterministic laws or the whims of physical chance events. His more introspective views, with his explanation of how individuals acquire their personal world views, were influenced by Schopenhauer and by the Indian Vedanta philosophy. Here he was convinced of the following: whatever we know about the world, be this knowledge scientific or prescientific, is mirrored solely in our sensations, perceptions, memories, and, of course, our language. The latter acts as the big mediator and social communicator in our culture. To be conscious of something, then, means nothing other than something being mirrored in such a way.

In The *Position of Schrödinger in the Einstein-Bohr Debate*, Ulrich Röseberg analyzes a great number of as yet unpublished letters which were exchanged between Schrödinger and other scientists. This analysis shows that the foundation of quantum physics and the new scientific world-view were actually more a result of a creative cooperation than of the usual competitive fight among scientists.

Schrödinger, when asked by a Viennese philosopher about his position and attitude in the Einstein-Bohr dispute, answered: "I tried to cooperate with them." He liked to compare his position with Plato's stance in the Greek dispute about the interpretation of reality. For Plato, this dispute resembled the mythical gigantomachy, the deadly fight between the giants and the gods. Like the Einstein-Bohr dispute, the Greek philosophical dispute was fought between two schools, the Eleatic metaphysical philosophers (Parmenides, Zeno, Pythagoras) and the Milesian empirical philosophical school (Heraclitus, the sophists). The latter - the so-called "panmovers" - conceived of the universe as eternally changing, as being in a constant flux. Only change was real for the panmovers. Their position is reflected in an idea which is gaining ground today even in the social sciences, namely that nature is largely governed by flux, disturbances and chance. For the Eleatic philosophers, change did not really exist; the "real" reality behind change was seen as unchangeable. In his witty way, Schrödinger compared himself to Plato, placing himself between the two groups, the group around Einstein, Rosen, and Podolsky on the one hand, and the Copenhagen physicists Bohr, Born, and Heisenberg, on the other.

But whoever thinks that this debate is a thing of the past, to the history of the thirties, is mistaken. Today, Bohm, Everett, Wheeler with his many-world interpretation of quantum physics, Tipler and Barrow side with the first group (see Kanitscheider's paper in this volume); Wigner, Penrose, just to mention two, have sided with the second. Some physicists have even managed to shift from the Einstein group to the Copenhagen school, as, for instance, the Russian physicist Blochinzev.

Thus, the paradox of the Einstein-Rosen-Podolsky experiment and Schrödinger's cat remain a challenge. We have yet to answer the following questions: Does reality exist independently of the observer and experimenter and their consciousness. Or does it not exist at all until observed?

In his article, Röseberg reveals an astonishing fact: The explosive development of quantum physics in Schrödinger's time was based on the close cooperation of a whole group of scientists, despite their divergent opinions. Thus, the new physics, the new revolutionary knowledge and world view were born from cooperation, something unheard of before in history of science. It may be seen as significant for our culture that cooperation effected more than the usual fierce, competitive fight for personal superiority or for the glory of academic institutions, as described by Merton.

Röseberg bases his "cooperative" thesis on the exchange of scholarly letters between the concerned parties. In his article, he cites excerpts from them, many of them translated into English for the first time. Schrödinger's contributions to science as documented in his correspondence demonstrate what scientific cooperation can achieve. The history and the course of the debate discussed above is revealed in the letters written by Schrödinger and M. Planck, N. Bohr, M. Born, W. Heisenberg, P. Jordan, A. Einstein, von Laue, Rosen and Podolsky, W. Pauli, B. Russell, and K. Popper. The letters clearly show the individual contributions, the different emphases as expressed in personal philosophical positions, the quite divergent assumptions in a debate out of which our view of the world today was born.

H. Poser's article, *The Notion of Consciousness in Schrödiger's Philosophy of Nature*, reconstructs Schrödinger's exposition of how the scientific world view is integrated into our own view of the world. This is the other side of Schrödinger's epistemology, his introspective view of how we synthesize our individual holistic outlook and how the world of scientists becomes my world or my picture of it. Poser contrasts Schrödinger's point of view not only - as Schrödinger did himself - with contemporary antagonists, but also with recent theories of perception and of knowledge and philosophical doctrines, including the New Age philosophy.

According to Poser's meticulous investigation, Schrödinger's central idea is a new concept of individual consciousness, a concept which was partly influenced by his quantum-physical holism, partly by his new interpretation of the role and function of the genes, and partly by Far-Eastern views about the unity of the world and the I.

In his reconstruction and critique of Schrödinger's new epistemology, Poser concentrates on the relation between the knowing individual subject and the object of our individual knowledge. In Schrödinger's holism, as opposed to dualism, they cannot fall apart, as in Kantian epistemology. Rather, they are cognitively connected in a complementary holistic unit; they are linked together through the functions of the genome, the sensory and nervous system, and the human brain. After having rejected in his wave mechanics the idea of an inalterable substance residing in the atoms, Schrödinger adamantly denied the existence of an unchanging substance as the substratum of the I, which is traditionally known as the immortal soul. However, he grants, in keeping with Indian Philosophy, that each individual ego contributes to our common world view, to our cultural knowledge in a way which we may call evolutionary. While we are growing up, we acquire cultural knowledge - and later we contribute our life-time experiences to a common pool, i.e., our world view which tradition preserves over generations.

By necessity, Schrödinger touches on the problem of metaphysics and nonscientific experience. For Schrödinger, anything concerned with the supraphysical is metaphysics. Metaphysics transcends science; it may complement our world view, provided it does not contradict science. Metaphysics should fill the gaps in our world view which are not yet filled by science. This is one of the reasons why Schrödinger used Indian Vedanta philosophy to support his scientific views and to make them more plausible, as, for example, in his theory of how we gain personal knowledge of our world and then integrate it into our individual view of the world.

He actually unravelled the problem of the ego and consciousness with only minimum recourse to metaphysics (if one assumes Vedanta philosophy to be metaphysics). As Elkana writes, B. Bertotti called Schrödinger's metaphysics a rational mysticism. Schrödinger started from his quantum-physical explanations of the interaction of the genes with the body and the brain. Based on this explanation, he developed a full theory of the cognitive mental process of how the 'I' synthesizes our knowledge of our world into an individual world view. He used a physiological cognitive models to explain the interaction of brain and 'I' and the concept of cultural-linguistic mediation to clarify the interaction of the ego and the social community.

According to Schrödinger's holism and his cognitive explanation of our individual process of knowing, the external world - even though it exists out there independently - and my consciousness cannot be separated. First, we are physically connected with the external world through our sensations. But the external world appears to me and to my consciousness only as my sensations, perceptions, memories, in brief: as my mentifacts. Second, since only single data, individual information, enter via our sense organs, our consciousness, the full inner picture has to be constructed and synthesized by my cognitive, mental functions. These perceptions, sensations and memories can only be connected and ordered through our imagination and our memory: this is our inner world-view. It makes no difference whether we do this with the help of scientific knowledge, mystical insights or even prescientific beliefs, as long as we verify or confirm our inner picture through empirical reality. Since this comes close to the views of Indian Vedanta philosophy, Schrödinger deliberately uses its liberal epistemology to support his introspective model of cognition.

In both versions, my personal world view exists only as my inner mental construction, as representation or combination of sensations, perceptions, memories, concepts, and mentifacts, ordered and woven to form a unity through the activity of the 'I'. Schrödinger shows why Western philosophers encounter almost insurmountable problems when dealing with such topics. If my world is my mind's individual internal representation of the external world, then it is finite; the world ends for me with my death. Therefore, for Schrödinger and the Vedanta philosophy, my I, or the assumed substrate called "soul", cannot be immortal. Since my world picture is my internal, unique, singular view of the world, we cannot say much - and perhaps nothing - about the world pictures in another 'I' or consciousness.

According to Schrödinger, there is no plural for expressing the uniqueness of my 'I' and for my consciousness. Poser explains in detail why Schrödinger forbids us to speak about the 'I' or consciousness in plural. But then how does Schrödinger explain how common and individual world view interrelate? In Vedanta philosophy, each single individual is just slightly different from the other individuals. Likewise, Schrödinger thinks that the similarity between the individual human genome (the genes in the gene pool) guarantees that the different pictures of reality in each I or individual are similar, in the sense that they are only slightly different facets or different copies of the same external objects. To use Schrödinger's metaphorical comparison: the cognitive functions of the 'I' make it resemble a hidden canvas on which all memories can be projected. To elaborate Schrödinger's metaphor: The cognitive process is like filming the same objects, the same scenes with (slightly) different cameras and then viewing each scene we have filmed. I can use language to describe my internal word-view to others; but I cannot invite someone else to view it on my canvas. This internal "viewing" of memories and even their creative combinations is what constitutes our consciousness. This 'I' or consciousness cannot be directly transferred to others. This is only possible indirectly via language in a way that does not interfere with reality - as Wigner assumes - just as pictures in a film cannot directly interfere with reality.

This is how Schrödinger reconciles the individual world view with the scientific one. Both should be subjected to verification procedures. His solution avoids many paradoxes, one of them being the paradox of free will. According to Schrödinger, moral laws are the expression of mankind's biological, evolutionary transformation from egoistic to altruistic, cooperative beings.

Poser succeeds in drawing a clear line between, for example, Capra's fashionable New Age philosophy and Schrödinger's philosophy of science. The difference lies in the discrepancy between a philosophy of science which includes science and a philosophy of science "without science."

Götschl's contribution *The Study of Nature and the Emergence of World View - Philosophical Reflections on Schrödinger's Approach* reconstructs Schrödinger's new evolutionary concept of world view, which ultimately brought the physicist in opposition to the offical Nazi-philosophy of his time.

Götschl's reconstruction begins with the drastic change in man's view of the universe and his world which shook physics and philosophy in the thirties. It was the disappearance of the harmonious unity of man and world, the legacy of Greek and European thinking. The author sees the reevaluation and the transformation of the classical world view as being based on Schrödinger's new dynamic concept of nature. Here nature is understood as an evolutionary process, in which man is no longer an onlooker but an active player.

While Schrödinger's new world view was no longer compatible with the traditional static understanding of the world, it was also not sufficiently reflected by our modern, scientific knowledge of nature. Götschl describes Schrödinger's lifelong search for a new "metaphysics" and explains in detail why it was not to be found in either metaphysics in the traditional sense or in just scientific progress. With man and nature becoming emergent components in the self-organizing process of our culture, scientific cognition is central to Schrödinger's new world view. Any increase in knowledge of nature thus corresponds to an increase in human self-understanding with cognitive science becoming the most powerful weapon against any kind of metaphysical, speculative world view. According to Götschl, Schrödinger's reevaluation of traditional metaphysics can be seen as a modern synthesis of knowledge and man's understanding of himself. Scientific knowledge is based on invariance or symmetry, not on an idealistic, epistemological concept of subject vis-à-vis inalterable objects.

According to Schrödinger, the scientific progress in man's view of the universe and nature stipulated a new dynamic unity between nature and culture. In his world view man is part of nature and cannot be reduced to physics or chemistry, since both cultural evolution and biological evolution are parts of a common, irreversible and non-linear evolution. In the new philosophy and in Schrödinger's new world view, subject and object relate to each other in an evolutionary way. Here the physicist anticipated the central idea of today's cognitive science: there is no longer a separation between matter and mind, subject and object, brain and consciousness.

Man is also no longer seen as an isolated part of the world. Like nature he is guided - albeit in an indeterministic way - through a process of universal evolution. Man and nature do not coexist but co-evolve. In this sense we have to understand Schrödinger's bold claim that scientific cognition of the world becomes self-knowledge. The integration of knowledge of nature into the dynamic network of a dynamic culture culminates in ethics and morality, in spite of the fact that morality cannot be reduced to science. The thinking subject is no longer just a linear extension of scientific reasoning but an active part of the approximation between culture and nature.

CONTRIBUTORS

Kurt Baumann

Kurt Baumann is Professor of Theoretical Physics at the University of Graz. He is co-author of a book on *The Interpretations of Quantum Theory* (1984).

Yehuda Elkana

Professor Elkana is Director of, and teaches at the Cohn Institute for the History and Philosophy of Science and Ideas. He is also the Director of the Van Leer Jerusalem Institute. His publications include the following books: *The Discovery of the Conservation of Energy* (1974), *Albert Einstein. Historical and Cultural Perspectives* (1982, edited by Gerald Holton and Yehuda Elkana), *Anthropologie der Erkenntnis. Die Entwicklung des Wissens als Episches Theater einer listigen Vernunft* (1986).

Johann Götschl

Johann Götschl is head of the Ludwig Boltzmann Institute for Science of Research and Professor of Philosophy of Science at the University of Graz. His previous writings include *Philosophical and Scientific Conceptions of Nature and the Place of Responsibility* (1988) and *Hypercritical Physical Realism and the Categorial Changes in the Subject-Object-Relations* (1991).

Leopold Halpern

Leopold Halpern is at present senior research physicist at Florida State University in Tallahassee. He was associated to Erwin Schrödinger from 1956-59 and to P.A.M. Dirac from 1974-84. His writings include: *On the Gravitational Radiation of Microscopic Systems* (1963), *The Measurement of Cosmological Variations of the Gravitational Constant* (1978), *Erwin Schrödinger's views on Gravitation during his last years in Vienna and some Research ensuing from it* (1987).

Herbert Hörz

Herbert Hörz is Professor of Philosophy of Science at the Institute of Philosophy of Berlin. His books include *Werner Heisenberg und die Philosophie* (1966), *Materiestruktur* (1971), *Zufall* (1980), *Wissenschaft als Prozeß* (1988), *Philosophie der Zeit* (1990).

Bernulf Kanitscheider

Bernulf Kanitscheider is Professor of Philosophy of Science at the University of Giessen. His books include *Philosophie und moderne Physik* (1979), *Kosmologie* (1984, second edition 1991) and *Das Weltbild Albert Einsteins* (1988).

Werner Leinfellner

Werner Leinfellner, University of Nebraska, Lincoln, U.S.A, and Technical University of Vienna, Austria. His books include: *Ontologie, Systemtheorie und Semantik* (1979), *Decision Theory and Social Ethics* (1979), *Einführung in die Wissenschafts- und Erkenntnistheorie* (1980). He is general editor of the Library "Theory and Decision" and founding editor of the journal "Theory and Decision".

Herbert Pietschmann

Herbert Pietschmann is Professor of Theoretical Physics at the University of Vienna. His books include: *Weak Interactions: Formulae, Results and Derivations* (1983), *Die Wahrheit liegt nicht in der Mitte* (1990).

Hans Poser

Hans Poser is Professor of Philosophy at the Technical University of Berlin. His books include *Zur Theorie der Modalbegriffe bei G. W. Leibniz* (1969); articles on philosophy of science and philosophy of mathematics.

Rupert Riedl

Rupert Riedl is Professor of Zoology at the University of Vienna. His books include: *Die Ordnung des Lebendigen* (1975), *Die Strategie der Genesis* (1976), *Biologie der Erkenntnis* (1980), *Evolution und Erkenntnis* (1982), *Biology of Knowledge. The Evolutionary Basis of Reason* (1984), *Die Spaltung des Weltbildes* (1985).

Ulrich Röseberg

Ulrich Röseberg is Professor, at present Visiting Fellow at the Center for Philosophy of Science at the University of Pittsburgh. His books include *Szenarium einer Revolution. Nichtrelativistische Quantenmechanik und philosophische Widerspruchsproblematik* (1984) and *Niels Bohr. Leben und Werk eines Atomphysikers* (1985).

Erhard Scheibe

Erhard Scheibe is Professor of Philosophy at the University of Heidelberg. His publications are on various topics in the Philosophy of Physics and include the two books *Die kontingenten Aussagen in der Physik* (1964) and *The Logical Analysis of Quantum Mechanics* (1973).

Franz M. Wuketits

Franz M. Wuketits is Professor of Philosophy of Science at the University of Vienna and Lecturer of Philosophy of Biology at the University of Graz. His books include *Zustand und Bewußtsein: Leben als biophilosophische Synthese* (1985), *Gene, Kultur und Moral* (1990), *Evolutionary Epistemology and Its Implications for Humankind* (1990).

ERWIN SCHRÖDINGER - A BIOGRAPHICAL SKETCH

JOHANN GÖTSCHL and GÜNTHER A. HÖFLER

Ludwig Boltzmann Institut für Wissenschaftsforschung und Institut für Philosophie der Universität Graz Mozartgasse 14, 8010 Graz, Austria

Erwin Schrödinger, who never liked chronologically structured biographies, only saw certain individual experiences as being significant. He was 72 when he presented an autobiographical overview that focuses on six periods of his life. The following sketch is thus in accordance with his wishes in that it does not attempt to give a complete account of his life and only concentrates on certain important aspects. In this first period, from 1887 to 1920, Schrödinger's special relationship with his father was most important during his childhood and teenage years. Rudolf Schrödinger was his intellectual mentor, someone he could confide in and also discuss every subject from school. From this period of his life Schrödinger remembers best the unforgettable talks he had with his father who had an amateur interest in botany. They discussed Darwin's *The Origin of Species*, a book Schrödinger remained enthusiastic about for the rest of his life. The nightly "talks" Schrodinger had with his deceased father whenever something extraordinary happened are illustrative of this very close relationship. Schrödinger grew up under exceptionally favorable circumstances. Being the only child in a well-to-do family, he had lacked nothing that could be beneficial to his intellectual development from a private tutor at elementary school level to extensive travel and regular visits to the theatre. His mother, who came from England, and her sister taught him to speak excellent English when he was a young boy. (This may have influenced his later decision to go to Oxford and Dublin.) Schrödinger was described as having been a sensitive but bright and vivacious child. It is thus not surprising that he had happy memories of his childhood and youth. His grammar school years at the Wiener Akademisches Gymnasium were also rather carefree and rewarding since apart from his conspicuous gift for mathematics he was also above average in all of the other subjects. Schrödinger also excelled as a student at the University of Vienna (1906-1910). His mathematics professors Wirtinger and Gustav Kohn and the experimental physicist, Franz Exner made an especially strong impression on him. However, it was the theoretical physicist, Fritz Hasenöhrl who had the strongest influence on him. Hasenöhrl who succeeded Ludwig Boltzmann, who died in 1906, fascinated his audience with his brilliant inaugural lecture. His acquaintance with Hasenöhrl was so important to Schrödinger that he even equated it with his father's intellectual influence. Apart from his specialized fields of study, philosophy appealed to him especially. The interrelationship of philosophy and physics was to become one of the most striking focus of his research. Schrödinger was also interested in literature which was reflected in lifelong admiration for Franz Grillparzer. Another important aspect of Schrödinger's years as a student is his closest friend Fränzel Frimmel von Traisenau, whom he mentions at the beginning of his autobiographical sketch. He and Frimmel, a student of biology, used to discuss fundamental philosophical issues at length, in particular, Darwin's theory of evolution and Richard Semon's *Mneme as the Conservative Principle*. Schrödinger was

J. Götschl (ed.), Erwin Schrödinger's World View, 19–22.

especially influenced by the former, evident in his negative attitude towards religion, which he never considered changing. After completing his studies, he was to do one year of military service (1910/11). Because of this he did not become assistant professor under Hasenöhrl but under Exner in 1911. In a sense he regarded this as an advantage since his background was mainly theoretical and he now had the opportunity to gain experience in experimental physics. This period of his life was suddenly interrupted in July 1914 when Schrödinger was conscipted immediately after the outbreak of World War 1. He served in the Austrian Imperial and Royal Army as officer until the end of the war. He saw action mainly on the Italian front where he served as an artilleryman and later as a metereologist. It was in Prosecco, Italy that he became familiar with Einstein's theory of relativity which was developed in 1915. The post-war period was anything but rosy one for the Schrödinger family. They had to give up their factory, and both money and food were scarce. The professional career of the young scholar, who had been offered the position of Associate Professor for Theoretical Physics in Czernowitz, also came to a halt at that time since that city no longer belonged to Austria. His father's death in 1919 coincided with the rise in inflation, which rendered the bank capital held by his father valueless. Schrödinger's situation improved, however, when he was offered an assistant professorship under Max Wien in Jena. His acceptance of this offer in 1920 after his marriage marks the beginning of the second period in Schrödinger's life, a period he called "the first period of travel."

In the winter term of 1920/21 he lectured at the Technical University of Stuttgart. In the summer term of 1921, when he was professor at the University of Breslau, he was offered the chair held by Alfred Einstein at the University of Zurich between 1909 and 1911. In December 1922 he delivered his inaugural lecture in Zurich on the topic "What is a Law of Nature?" The fact that this lecture was scheduled one and half years after his appointment probably had something to do with the psychological crisis Schrödinger was going through at the time. (This crisis may have contributed to his illness (tuberculosis) later in life.) Towards the end of 1925, during a stay at Arosa, he formalized his revolutionary theory of wave mechanics, which he published in 1926. His formulation of the wave equation met with lively response from the scientific community; he was invited to go on a two-month lecture tour through the United States in the winter of 1926/27 and offered a professorship in Berlin as successor of Max Planck. In his autobiography this move to Berlin marks the beginning of the third period of his life, the so-called "Berlin period of learning and teaching." In retrospect he saw this period as a very good time in his life, even though it came to an abrupt end after the summer term of 1922 when the Nazis seized power. During those six years he became acquainted with Einstein, Planck, Hertz and Otto Hahn. He got along very well with all of them and there was a mutual feeling of respect. In the spring of 1935 Schrödinger, while on granted leave as a guest professor in Oxford, informed his department in Berlin that he did not intend to resume his duties in Berlin. His resignaton was not formally accepted until March 31, 1935. A detailed biography* would elaborate on the summer of 1933 as a time of tension, uncertainty and sorrow for Schrödinger. It was, however, also a time marked by happy events in his private life (his relationships with women) - something he does not deal with in his autobiography.

Schrödinger refers to the fourth period of his life as his "second period of travel." At the beginning of October 1933 he came to Oxford at the invitation of F. A. Lindemann, with whom he had reached an agreement at Lake Garda in the summer. During his first stay in Oxford (October 1933 - September 1936), where he was a fellow at Magdalen College, he was awarded the Nobel prize. He received it on the 10th of December 1933 - a date he fails to mention in his autobiography. He also makes no mention of the three lecture tours he did at the time. The first one took him to Princeton in the spring of 1934, and the subsequent ones to Spain in August 1934 and from March to May 1935. Due to his exceptional gift for languages Schrödinger was able to deliver his lectures in Spanish. Soon after his six-week sojourn in Princeton he was offered a professorship at the University of Princeton which he did not accept. In 1936, however, given the choice between a professorial chair in Edinburgh and one in Graz, he decided in favor of Graz. Before that, in the summer of 1935, he had published a much-discussed paper entitled "The Current Situation in Quantum Mechanics" in three parts, which founded the fame of the so frequently cited cat. There has been much speculation on what finally made him decide to return to Austria, despite the precarious political situation there at the time. This question has yet to be fully resolved. Schrödinger only refers to his decision as an "unparalleled blunder." While professor in Graz (winter term 1936/37) he was also honorary professor at the University of Vienna, where he held a special two-hour lecture every two weeks.

His second flight from Hitler in 1938 was preceded by a Gestapo interrogation, which Schrödinger saw as alarming signal. Immediately afterwards he left Austria for Rome, where he decided to accept a position at the Dublin Institute for Advanced Studies, which had been offered to him by the Irish Prime Minister De Valera. He had also received offers from Brooklyn and Turkey. As the Institute in Dublin had not been fully established at the time, he returned to Oxford for two and a half months. In mid-December 1938, he accepted a guest professorship in Belgium (University of Ghent) where he stayed until autumn of the following year. He described the welcome in Ghent as warm and his working conditions there as highly satisfactory. After he and his wife had been smuggled through England as "enemy-aliens" he arrived in Dublin on the 7th of October 1939. According to Schrödinger, his arrival constituted the beginning of the fifth period of his life, which he refers to as the "long exile." He describes the sixteen years he spent in Ireland as very good ones, not only because living in a neutral country meant he was unaffected by the war but also because of the agreeable conditions that had been created for him as a scientist. The Institute included a "School of Theoretical Physics" where he enjoyed full freedom as a teacher and researcher without having to teach. In this congenial environment a very prolific period of his career developed. In 1943, he began working on "The General Unitary Theory of Physical Fields" which was a draft of a unified field theory. It should be noted here that apart from publishing such great works as "Statistical Thermodynamics" and the immensely popular and much-discussed *What is Life?* and *Nature and the Greeks*, he also found time to pursue his literary and artistic inclinations, such as poetry writing. Moreover, his gift for languages allowed him to translate Homer from ancient Greek into English and old Provencal poetry into German.

Schrödinger turned down a professorship offer he received from Austria in 1946 on account of the still unclear political situation in Central Europe. Also, no agreement could be reached at that time on the social security provisions for his wife. In 1950 and 1951 he participated at the European Forum in Alpbach and spent his holiday in Seefeld. In the winter term of 1950/51 he was guest professor at the University of Innsbruck. All of this may have made Schrödinger more inclined to return to his home country. In 1956 he accepted a professorship "ad personam" in Vienna, arriving here on the 5th of April after a short stay in Kitzbühel. Schrödinger sees this as the beginning of the sixth period of his life.

His inaugural lecture on "The Crisis of the Nuclear Concept" was enthusiastically received. He lectured for two and a half years in Vienna before illness forced him to give up teaching. After retiring from his chair at the end of September 1958, he spent the last years of his life more or less in seclusion. During this time he wrote his autobiographical sketch *My Life* as well as his philosophical will *My View of the World.* Schrödinger spent the last months of his life in rather poor health condition in Alpbach and Vienna, respectively. The letters he wrote to his friend, the poet Franz Theodor Csokor, reflect his suffering. On the 4th of January 1961, he died of "general arterioscleroisis." He was laid to rest in Alpbach.

(Acknowledgement: Our sincerest thanks go to Frau Hofrat Dr. Auguste Dick who provided us with material and useful advice.)

*) See Walter Moore, *Erwin Schrödinger - Life and Thought*, Cambridge University Press, Cambridge 1987, for an excellent, very comprehensive survey of Erwin Schrödinger's life and work.

PART I

PHYSICS IN THE PHILOSOPHICAL SYSTEM OF ERWIN SCHRÖDINGER

ERWIN SCHRÖDINGER AND THE PHILOSOPHY OF THE PHYSICISTS

ERHARD SCHEIBE
Philosophisches Seminar der Universität Heidelberg
Marsiliusplatz 1, 6900 Heidelberg, Germany

I.

Erwin Schrödinger whose hundredth birthday we celebrated not long ago was awarded the Nobel prize for physics in 1933. He was honored "for the discovery of new fruitful forms of atomic theory" that he had made several years earlier and first published in his famous papers "Quantisierung als Eigenwertproblem."[1] The community of physicists decided to call the basic dynamic equation of quantum mechanics the "Schrödinger equation". The view, associated with this equation, that quantum mechanics is a theory of the temporal changes of state in a physical system is known as the "Schrödinger picture". All this seems to show that Schrödinger was deeply involved with the new atomic theory - the most important advancement in physics in this century. On the other hand, it is a well-known fact that he did *not* accept the orthodox interpretation of quantum mechanics. And he was not the only one to find himself in this situation. Already Planck and Einstein were awarded the Nobel prize for achievements marking the very beginnings of quantum theory. Yet both men were not satisfied with the interpretation of a theory which was viewed as definitive by the majority of physicists. This coincidence indicates that something peculiar was going on - something transcending physics proper but still belonging to it in a broader sense.

What was at issue here is easily related in general terms. Physics is grounded not only on the laws of nature that it attempts to discover but also on tacit assumptions of a more general kind. There is nothing unusual about this, and it is something most physicists are unaware of. The special nature of physics, the exception, only results from physicists *becoming aware* of one or the other tacit assumption. A change usually occurs when fundamental difficulties arise in the discipline, as was the case in atomic physics at the beginning of the century. It became so difficult to understand the structure of the atom and the emission and absorption processes of light as well as the strange dualistic behavior of free radiation fields and particles that it was unlikely that a solution could be found within the bounds of classical physics. The behavior of matter on an atomic scale motivated instead a radical revision of physical thought that would break with some of the epistemological, ontological and even logical assumptions underlying traditional physics. Of course only physicists knew which modifications were to be made, and correspondingly, which new foundation physics would obtain. They were thus forced by the very development of their discipline to *philosophize*, not in the trivial sense in which everybody has his world view but in order to overcome a serious foundational crisis in their discipline.

If in the title of this paper I allude to the "philosophy of the physicists" it is not a particular philosophical doctrine that I am referring to - no more than such is meant when we speak of the "philosophy of the Greeks". What I am referring to is the intellectual struggle of physicists with

J. Götschl (ed.), Erwin Schrödinger's World View, 25–34.

fundamental issues, including space and time, as they arose during the first half of this century. Though it is widely known that men like Bohr, Einstein, Heisenberg, Born, Jordan, Planck and, last but not least, Schrödinger were involved in this struggle and though many papers dealing with special issues in this field were published, the history of the philosophy of physicists has yet to be written. I cannot, of course, do this in this paper. But I would like to present the following considerations focusing primarily on Schrödinger within this larger context. For this physicist movement did in fact contribute significantly to the philosophy of science. Its occasional dilettantism is on the whole less irritating than much of the philosophical work done in the field by professionals.

Schrödinger's contribution is governed - as I said in the beginning - by his basic opposition to the orthodox view of quantum mechanics. From what I then said, it follows that this opposition is located in that domain between physics in the narrower and physics in the wider sense, where even this discipline with its high level of unity has gaps that can be filled only by more or less philosophical assumptions. I say "more or less" because, as we shall soon see, though Schrödinger was deeply convinced of the importance of a genuine philosophical understanding of the world, he expressed doubts as to the philosophical relevance of the problems posed by quantum mechanics and their orthodox solution. Therefore, if one wants to ask, as I will do in the following, what Schrödinger's opposition was like and what led up to it, one has to traverse the vast field between quantum jumps and Vedanta. The reader will understand that this can only be done in "quantum jumps."

II.

In an essay, in which he compares the philosophy of Bohr, Heisenberg and Schrödinger, Abner Shimony says he believes that Schrödinger "was the most remarkable philosopher among the physicists of our century."[2] I have no immediate objections to this statement. Is, however, this supposed to mean that Schrödinger has an *important* philosophical message *in connection* with the foundational crisis of physics mentioned before? Is this his message to us - a message comparable with Bohr's idea of complementarity or Heisenberg's modal interpretation of quantum mechanical reality? To answer this question let me first stake out the field with two cornerstones that we can rely on. For one thing, Schrödinger saw the development of physics beginning at the turn of the century as being in a state of *genuine crisis*. In *Nature and the Greeks*[3] he writes:

> My point is this: The modern development, which those who have brought it to the fore are yet far from really understanding, has intruded into the relatively simple scheme of physics which towards the end of the nineteenth century looked fairly stabilized. This intrusion has, in a way, overthrown what had been built on the foundations laid in the seventeenth century, mainly by Galileo, Huygens and Newton. The very foundations were shaken."

On the other hand, both Schrödinger and Dirac, with whom he won the Nobel Prize, were convinced that the crisis could not be overcome by the quantum mechanics of the late twenties. Schrödinger ironically compared quantum jumps with the epicycles of Ptolemean astronomy[4] and he belittled the "pulling down of the frontier between observer and observed" as being "a much overrated aspect without profound significance."[5]

These quotations show that while Schrödinger took the crisis of physics seriously, he seems to have been inclined to locate a possible solution in the philosophical considerations of physics *closer to physics* than to philosophy but by no means closer to the orthodox philosophy of quantum mechanics. We can learn more about his thinking when we examine his general view on the relationship between science and philosophy. From his Berlin period there is the following remark (one I would describe as conventional) that he made in connection with the "present raise of the causality problem":

The old alliance between philosophy and natural science is formed anew...The further natural science progresses the less it can do so without philosophical critique...[6]

Two remarks he made one year before his death are even more revealing. In a letter[7] he writes:

It would be...quite wrong to say that the development of science as a whole has no influence on philosophical thought; but never in the way that one clear-cut...theorem of one special science has a logically irrefutable consequence in philosophy.

Schrödinger also expresses himself very clearly in the preface to his philosophical legacy *Meine Weltansicht*.[8]

Here he discloses to the reader not only that as a young man, he was on the verge of doing physics just for his livelihood whereas in his spare time, he would do philosophy as his proper concern. He also says in well-weighed words that now in the old man's book "he would nowhere talk about acausality, wave mechanics, uncertainty relations, complementarity, expanding universe, continuous creation, etc." because "these things seem to have much less to do with the philosophical world view than is popular today."[9]

Here again he is speaking in clear terms. The reader is told that the foundational crisis of physics which affected Schrödinger *as a physicist* hardly touched his private philosophical world view. We will now seek to gain more than just a general perspective by briefly examining this view. Schrödinger must have been an introvert for whom it was natural to interpret the world as it appeared to him mentally and emotionally. Of course, this did not make him a solipsist: there were those other individuals, each with his own inner life. There was also the possibility of communicating with our fellow man, this communication being of particular importance in the natural sciences. To Schrödinger, the fact that many independently existing human beings can come to believe that they live in one common world was *the* great miracle of the world. His main philosophical problem therefore was: How, if at all, can we explain this miracle?

The usual explanation, which Schrödinger assumed that also the majority of his colleagues endorsed, is that there is a real world of objects out there to which, among many other things, our own bodies belong. Schrödinger did not accept this view.[10]

He found it naïve and not sufficient to explain the miracle. Above all he has emphasized that it is at least as metaphysical and as mystical as the view that he accepted. It is metaphysical because this supposed world of real objects would by definition not be observable. And it is mystical because the alleged causal nexus connecting real objets with perceptions of conscious beings would indeed be entirely mysterious.

However, in the last resort it was moral and religious grounds that led Schrödinger to solve this main problem with an idealistic monism. Influenced by philosophers like Spinoza and Schopenhauer and, on the other hand, by the Indian Vedanta, he believed in a universal spirit of which the various individuals were only aspects and whose unity would guarantee the community of common experience that was to be explained. Nature, however, as we generally come to know through our senses, has no detached existence. It is made from the same stuff as that universal spirit, i.e., our sensations. Schrödinger on the great universe which cosmology presents to us and the romance of a world that eventually produces brains so that it can be understood: "To me personally all this is *maya*, even if very lawful and interesting *maya*."[11] What really counts in view of such a beautiful delusion is the moral content of the doctrine of the identity, the *numerical* identity of the self with the other selves and the religious consolation to partake in a timeless, eternal existence.

III.

Once one has realized that this was Schrödinger's world view and that he took it at least as seriously as physics, it becomes clear what the difficulties of his discipline could ultimately mean to him. On the other hand, Schrödinger was not the man to use eastern wisdom to make his western life easy. This could be seen as the *basic contradiction* of his thinking: on the one hand, his affinity to eastern thought, which, when established as a religious form of life, has prevented a scientific world view from being established, and on other hand, his affiliation to western culture where science had been created and had developed - also assisted by Schrödinger - to the sophistication of wave mechanics. It was to be expected that Schrödinger wanted to find out in what *particular* sense the scientific world view was to be able to endure this tension. And to this end he made considerable intellectual efforts by going back to the sources - to Greek philosophy.[12] Whether his clarification is justifiable from the viewpoint of the historian need less concern us than what it led up to. It resulted in the view that the scientific world view is based on the realization of two postulates: the postulate of comprehensibility of nature and the postulate of its objectification. Schrödinger also pointed to a connection: the postulate of objectification has to be satisfied *in order* for nature to be understood. Of course, one must first explain what these postulates are all about. But already at this point it should be said how matter will evolve in view of our main problem. Schrödinger reached the conclusion that quantum mechanics violates *both* postulates, and it is for this reason that he criticized it. He did this *although* 1) he saw the restrictions that were imposed on the scientific world view precisely by the fulfillment of these postulates, 2) felt that they were restrictions affecting his *own* monistic world view and 3) again and again was on the verge of realizing that quantum theory could eventually loosen the restrictions a bit.

Schrödinger's assumption that nature can be *understood* and his postulate that a physical theory would allow a maximum of understanding, belong to the tradition of understanding in terms of pictures or models that dominated physics for sometime and can be traced back far into the 19th century.[13] Speaking in terms of *pictures* or *models* made it possible for physicists to turn against the rather gross dichotomy of false and true statements. Realizing that truth is rarely ever attained one had to look for something that mitigated the only alternative of being false. One came to say instead that physical pictures do more or less justice to the facts or that they are more or less useful. The question, which criteria could be accepted for ever better adaptation or what pictures were to be admitted at all, was not discussed for a long time. Even Schrödinger took up this question only when quantum theory made it evident that the kind of pictures used in classical physics led to serious trouble if applied on the atomic scale.

Schrödinger clearly saw that the use of pictures in physics cannot be specified in exact terms. He attempted to describe the assumption of comprehensibility as a solution situated between animism and positivism, animism being the archaic paradigm of understanding, Schrödinger considered worth mentioning in view of the revival of the causality debate. He wanted to remind people that any attempt to disclose the nature of causality would inevitably end up in primitive animism. It was hopeless to attain a causal understanding of nature. On the other hand, positivism was too modest. Understanding can only be attained if our pictures of nature show features of wholeness and "gestalt". This, however, never obtains if one follows positivism and restricts science to the observable, to sense perception or the like. With this restriction we sacrifice objectification and no coherent picture emerges. All connections are lost, and science is without theoretical orientation. It is here where the aforementioned connection between the two postulates becomes particularly perspicuous: We have to objectify in order to understand. The pictures are the end, not the means. As Schrödinger put it:

> As I find it, in physics...the appreciated result of our endeavors is a clearly drawn, pictorial overall view of the object investigated, completely understood in its inner connections. (These connections) would be entirely destroyed if we were to scruple about truthfulness and formulate all our statements in such a manner that their direct relation to sense perception is displayed.[14]

Objectification means above all the elimination of human consciousness from the world view of physics. Even for Schrödinger the world of real objects, rejected as a metaphysical entity, was to be accepted as a fiction. However, the elimination of consciousness was more a matter of fate than a postulate: Strictly speaking we *cannot*, not even outside the world of science, objectify consciousness. We do not find it in our world view "because it ... is this world view."[15] But there are things related to consciousness and less comprehensible: there are its activities and abilities, there are sensations and memory and the like. With respect to this wider range, objectification means to obtain pictures that describe nature without giving any hints as to how we came to know all this. For instance, the measuring instruments would not appear in these pictures in their functioning *as* measuring instruments. And not even the question of measurability would be raised by the pictures.

> Hardly any physicist of the classical period - says Schrödinger - inventing a model has had the imprudence to believe that its variables were measurable.[16]

Physics has to describe nature itself, not its relation to a knowing subject.

IV.

We are now ready to go one step further - perhaps the last practicable one - in answering our main question: what made Schrödinger oppose the orthodox interpretation of quantum mechanics? So far we have reduced this question to the following one: In what respect does quantum mechanics violate the postulates of comprehensibility and objectification? As regards *comprehensibility* it is in order to touch upon two subjects: determinism and description of space and time. According to the orthodox view, quantum mechanics is an essentially indeterministic, irreducibly probabilistic theory. Some would even declare this feature as being the *major* one that distinguishes quantum theory from classical physics (including relativity theory!). It is therefore worth mentioning that Schrödinger was far from putting indeterminism on his index of quantum theory. He was an admirer of Boltzmann, the inventor of statistical physics. His own papers on this topic fill one of the four volumes of his *Collected Works*. Once he had developed the view, which was summed up in the preceding section, statistical mechanics became the paradigm of understandible physics for him. Even in his last published paper (1958) Schrödinger considered the possibility that energy might be a statistical concept like temperature or entropy.[17]

All this does not say anything definite about Schrödinger's views regarding indeterminism. There are, however, two arguments in favor of interdeterminism that were known *before* the advent of quantum mechanics. One of them, going back to Boltzmann and Poincaré, shows that even an arbitrarily small imprecision in our knowledge of the initial conditions of a mechanical system leads, in the majority of cases, to complete ignorance about those conditions at a later time. This argument has been used by Born for appeasement regarding quantum mechanical indeterminacy. Today it has become integrated into chaos research·[18] The second argument stems from Franz Exner, one of Schrödinger's teachers. This argument forms the core of the inaugural lecture Schrödinger held in 1922 in Zurich[19]. His argument runs as follows: Determinism on a molecular level is not necessary. If macroscopic pressure can be explained statistically, why shouldn't this also apply to energy? Determinism is a habit of thought that has "evolved in millenia of observing precisely those regularities which we know today *not as causal* but as *directly statistical regularities*." Moreover, we know about this macroscopic regularity that "it would exist even if the course of every individual molecular process were decided by throwing dice." Therefore, nothing forces us to assume molecular determinism, and this, if anything, shows the power of statistical explanation. There are even arguments that make this determinism improbable. For it would lead to the following schizophrenic situation:

"In the realm of appearances, clear comprehensibility - behind it a dark, eternally incomprehensible power, a mysterious "must"...Such a double foundation of regularity in nature is improbable by itself. *The burden of proof lies on the advocates, not the sceptics of absolute causality*. For to doubt it is much *more natural* today."

Did this seem more natural five years later when it became clear what *quantum mechanical* interdeterminism looks like? Schrödinger never denied the heritage of Boltzmann and Exner. He dealt with the issue even after 1927. But his remarks - so it seems - became somewhat more cautious. Already in 1927 he presented the following interesting argument:

I shrink back from this conception ... because one should demand of a theory which postulates an absolute, primary probability...that at this price it should free us from the old 'ergodic difficulties' and make us understand the irreversibility of natural processes without further assumptions.[20]

Presumably it was not so much the primary probabilities *by themselves* Schrödinger shunned but the *reason* why an irreducible indeterminism had become necessary. The reason was the apparent impossibility of describing nature on the atomic scale in the classical way, i.e., by giving an account of what is actually going on at every point of a continuously connected region of space-time. Schrödinger realized that the phenomena embraced by the so-called wave-particle dualism presented a serious difficulty that somehow had to be overcome. But he was not ready to accept the orthodox solution in *this* respect: The renouncement of the classical spacetime description - the intrusion of discontinuity, of quantum jumps and the new kind of identity of particles - destroyed the ideal of nature as something understandable.

"There are - Schrödinger says - as it were, gaps in our picture."[21] However, possibly with the exception of a period around 1930, he firmly believed that the gaps could be eliminated. Already in one of his original papers on wave mechanics, Schrödinger speculated whether the processes in the atom can be integrated into the space-time form of our thinking. In an almost Wittgensteinian manner he says:

> From a philosophical point of view I would consider a definite decision in this sense to be a complete surrender. *For we cannot really alter the forms of thinking, and whatever we cannot understand within them we cannot understand at all.* There are such things, but I do not believe that atomic structure belongs to them.[22]

In a similar vein, in his last paper of 1958 he emphasizes:

> We do feel a yearning for a complete description of the material world in space and time, and we consider far from proven that this aim cannot be reached.[23]

Schrödinger's wave mechanics was his attempt to reach the aim. Wave mechanics represents the physical portion of the whole argumentation, and its importance was underlined by the Nobel prize. That the prize was divided between Schrödinger and Dirac, who favored particles, was an apt symbolization of the wave-particle dualism. Today the merits of wave mechanics are being assessed by quantum field theory. Indeed there was at least one good reason for Schrödinger to reject quantum jumps: his own equation, the basis of quantum mechanics, shows us that an atom once in an eigenstate of the energy has to stay there forever. The spontaneous emission and absorption of light is not explained by quantum mechanics. On the other hand, the theory which does explain it - quantum electrodynamics - certainly is a quantum theory. Therefore it is hard to reconcile with the postulate of comprehensibility and its realization in pure wave mechanics.[24]

V.

In conclusion we have to ask what has become of the postulate of *objectification.* It was already mentioned that this postulate was seen as serving the realization of the postulate of comprehensibility. As quantum theory violates the latter, it is to be expected that there are difficulties already with the former. And this has proven to be true. Moreover, the Copenhagen interpretation of quantum mechanics, the main target of Schrödinger's critique, has completely reversed the matter: the unattainability of the postulate of comprehensibility - the inapplicability of classical pictures of the atom - was *explained* by the impossibility of objectifying the processes: The observer had to be included more or less in the description of quantum phenomena. I say "more or less" because there is a certain latitude in assessing both the matter itself and its various interpretations. Correspondingly, Schrödinger shifts the focus of his critique back and forth. But it is always assumed that at some point one enters the sacrosanct district of consciousness, the atmosphere of subjectivity.[25]

At face value the Copenhagen view seems to open up possibilities undreamt-of. On account of his philosophical position, Schrödinger was convinced of the narrowness of the outlook which science had imposed on itself by the fulfillment of the postulate of objectification. He expresses himself in clear terms:

...I consider science an integrating part of our endeavour to answer the one great philosophical question which embraces all others,...*who are we*? And more than that: I consider this not only *one* of the tasks, but *the* task, of science, the only one that really counts.[26]

Consequently, when quantum theory seemed to force a renouncement of the objective description of nature Schrödinger at first hesitated. In a lecture of 1930 he admits that it is

...a painful reduction of our claims to truth and clarity that our symbols and formulas and the pictures associated with them do not represent an object independent of the observer but only the relation subject:object. *But* - Schrödinger continues - *is not this relation the only reality we know of strictly speaking*? Does it not suffice if *it* finds fixed, clear and entirely unique expression ...? Why do we have to eliminate ourselves by all means ...?[27]

Schrödinger never pursued these questions from the same perspective. On the contrary: the door that was open for some time is definitely closed by 1935. In this year the now famous paper by Einstein, Podolsky and Rosen was published.[28] Moreover, it is Schrödinger to whom we are indebted for the first thorough analysis of the situation. He at once realized the devastating epistemological consequences of *inseparability*: To most quantum mechanical pure state descriptions of a system composed of two subsystems no (pure) state descriptions of the subsystems correspond:

I would not call that *one* but rather *the* characteristic trait of quantum mechanics, the one that enforces its entire departure from classical lines of thought.[29]

However, precisely this break was too radical for Schrödinger: his cat paradox probably has become the best known expression of this attitude.[30] In spite of the gravity of the case Schrödinger could not believe that the intrusion of the "observer" pointed in a direction where the deep philosophical problem of the relation between subject and object is hidden.

It is not very easy to say *why* I do not believe it. I feel a certain incongruity between the applied means and the problem to be solved.[31]

Schrödinger's argument seems to have been: measured against the infinite problem of consciousness, what we learn from quantum theory is too small to compensate for the loss with regard to the two epistemological postulates in question. Consequently, in his later work on the philosophical implications of quantum theory Schrödinger always presented the matter against the background of the classical modes of understanding, thereby trying to reduce it to absurdity. He rarely attempted to tackle the problem of quantum theory on the grounds of his own philosophical position.[32] Perhaps he somewhat thoughtlessly lost a chance that presented itself to him of all persons. Perhaps it was his own fault if in this matter not he himself but only his cat became famous.

NOTES

1 *Abhandlungen zur Wellenmechanik*. Leipzig: Joh. Ambr. Barth 1927; 2 ed. 1928. (Articles quoted without the author are by Schrödinger.)

2 A. Shimony, Reflections on the Philosophy of Bohr, Heisenberg and Schrödinger. In: *Physics, Philosophy and Psychoanalysis*. Essays in Honor of Adolf Grünbaum. ed. R. S. Cohen and L. Laudan. Dordrecht: Reidel 1983. p. 209-21. quoted from p. 215.

3 *Nature and the Greeks*. Cambridge: CUP 1954. quoted from p. 15.

4 "Are there Quantum Jumps?" 4.478-502. quoted from p. 481. (Works quoted in this manner are taken from *Gesammelte Abhandlungen*, ed. by Österreichische Akademie der Wissenschaften, Vienna 1984. The first figure refers to the volume, the next one(s) to the pages.)

5 op. cit., no. 3, p. 34.

6 Das Gesetz der Zufälle. 4.316-7. quoted from p. 317.

7 B. Bertotti, "The later work of E. Schrödinger", *Stud. Histo. Phil.Sci. 16* (1985) 83-100, quoted from p. 85.

8 *Meine Weltansicht*, Vienna: Paul Zsolnay 1961.

9 ibid., p. 115 ff.; also op. cit. no. 7, p. 94.

10 op. cit. no. 8, pp. 153 ff.

11 op. cit. no. 8, p. 175.

12 op. cit. no. 3; Die Besonderheit des Weltbilds der Naturwissenschaft, 4.409 - 53.

13 *Locus classicus* is H. Hertz, *Die Prinzipien der Mechanik*. Joh.Ambr. Barth: Leipzig 1894. In addition to Schrödinger's papers referred to no. 12, see also: Der erkenntnistheoretische Wert physikalischer Modellvorstellungen (1928), 4.288-94.

14 op. cit., no. 12, p. 425.

15 op. cit., no. 12, p. 443.

16 Die gegenwärtige Situation in der Quantenmechanik, 4.484-501, quoted from p. 486.

17 Might perhaps energy be a merely statistical concept? 1.502-10. This idea, formed already early by Schrödinger, belongs to a consistent wave mechanical interpretation of the microphysical world.

18 M. Smoluchovski, "Über den Begriff des Zufalls und den Ursprung der Wahrscheinlichkeitsgesetze in der Physik," *Naturwissenschaften 6* (1918), pp. 253-63; Max Born, "Ist die klassische Mechanik tatsächlich deterministisch?" *Phys.Blätter 11* (1955) pp. 49-54; Max Born and D. J. Hooton, "Statistical Dynamics of Multiple Periodic Systems," *Zeitschrift für Physik 142* (1955), pp. 201-18.

19 F. Exner, *Vorlesungen über die physikalischen Grundlagen der Naturwissenschaften*. Deuticke: Wien 1919. part IV, esp. pp. 691 ff. Was ist ein Naturgesetz? 4.295-7. I see Forman's description of the development, in which he includes also Weyl, Reichenbach and others, as being a turn towards acausality as misleading. See P. Forman, "Weimar Culture, Causality and Quantum Theory. 1918-1927," in: *Hist. Stud. in the Phys. Scis. 3* (1971) 1-115, esp. III.3.

20 Energieaustausch nach der Wellenmechanik, 3.267-79, quoted from p. 279. See also: Antrittsrede, 4.303-7; Das Gesetz der Zufälle, 4.316-7; Die Wandlung des physikalischen Weltbegriffs (1930). In: *Was ist ein Naturgesetz?* Oldenbourg: Munich 1987. p. 18-26, esp. p. 24; Über Indeterminismus in der Physik (together with: Ist die Naturwissenschaft milieubedingt?) Joh.Ambr.Barth: Leipzig 1932. 1-24.

21 *Science and Humanism. Physics in our Time*. Cambridge University Press 1961. p. 27.

22 "Quantisierung als Eigenwertsproblem II" (1926), 3.98-136. Quoted from p. 117f. Italics are mine.

23 op. cit. no. 17, p. 509.

24 For a recent vindication see J. Dorling: "Schrödinger's original interpretation of the Schrödinger equation: a rescue attempt." In: *Schrödinger*. Centenary celebration of a polymath. Ed. by C. W. Kilmister. CUP 1987. 10-40. See also L. Wessels: *Schrödinger's Interpretation of Wave Mechanics*. Dissertation Indiana University at Bloomington, 1975.

25 It is not always clear which view Schrödinger is attacking. He seldom uses names or gives references. On the other hand, the views of the members of the Copenhagen school diverge.
See op. cit. in no. 2 and E. Scheibe, "Die Kopenhagener Schule" In: *Klassiker der Naturphilosophie*. ed. by G. Böhme. Beck: Munich 1989, pp. 374-92.

26 op. cit., no. 21, p. 51.

27 op. cit., no. 20 ("Die Wandlung des physikalischen Weltbegriffs"), p. 26. This lecture, given in Munich 1930, was not published until after Schrödinger's death.

28 A. Einstein, B. Podolsky, N. Rosen, "Can the Quantum-mechanical Description of Physical Reality be Considered Complete?" *Phys. Rev. 47* (1935), pp. 777-80. Schrödinger's reaction was the paper referred to in no. 16.

29 "Discussion of Probability Relations between Separated Systems," 1.424-32, quoted from p. 424. See also: Probability Relations between Separated Systems, 1.433-9; The philosophy of Experiment, 4.558-68.

30 op. cit. no. 16, p. 489.

31 op. cit. no. 21, p. 51f.

32 See op. cit. no. 17, pp. 507ff.

HOLISTIC ASPECTS OF THE SCHRÖDINGER EQUATION

HERBERT PIETSCHMANN
Institut für Theoretische Physik der Universität Wien
Boltzmanngasse 5, 1090 Wien, Austria

1. THE HUMAN SIDE OF ERWIN SCHRÖDINGER

The following reflections will focus on some issues related to the interpretation of Schrödinger's wave equation. Since, of course, the emphasis will be on Schrödinger's role in this game, it is in order to describe his personality at least in a few words. In doing so I would like to quote from the lecture Schrödinger held on the occasion of his nomination as full professor. Schrödinger dedicated this lecture to his predecessor Michael Radakowic and one of his successors, Paul Urban, edited this lecture.[1] The topic of this opening lecture in 1936 was: "Die Grundidee der Wellenmechanik" (The basic idea of wave mechanics). At the very beginning Schrödinger says that Michael Radakowic's way of thinking and working reflected uncompromising clarity, well-structured ideas and a refutation of any unclear babble which may seem ingenious but in truth only admits lack of understanding.

Schrödinger goes on to claim that Radakowic had inherited the unconditional integrity of his research partly from his great teacher Ludwig Boltzmann. He said "let us try to honor the memory of both men by simply working and thinking together..."

By praising his predecessors Schrödinger also characterizes himself. But in his own thinking "unconditional truth" also meant thinking in "either-or" categories. It was his straight-forward nature that led him an admirably straight way amidst political confusion: but it also meant that he could never accept what historically turned out to be the right interpretation of his own main contribution: quantum mechanics. At the end of his opening lecture he said:

> Let us go back to the beginning - remembering our deceased predecessors - and in so doing let us be honest. The theory is not as simple as I have presented it here. I am not referring to difficulties in mathematical development but in connotation.
>
> The pure wave picture is as insufficient as a pure particle picture. The truth lies - in between? No. We do not know. It would be easy to simply beat around the bush. But I will avoid this in consideration of our predecessors.

Schrödinger would not tolerate contradictions even in the sense of complementarity. In his philosophical book *Meine Weltansicht*[2] he writes:

> It appears to me that these two statements are contradictory so that it cannot even be suspected that both are right.

It is a well-known fact that Schrödinger was supported by Albert Einstein who wrote to him in a letter of 1950:

J. Götschl (ed.), Erwin Schrödinger's World View, 35–39.

(Apart from Laue) you are the only one among the contemporary physicists to recognize that we cannot avoid postulating reality - if we are honest. Most of them don't even notice what a dangerous game they are playing with reality.

What kind of "dangerous game with reality" is Einstein referring to? Wolfgang Pauli described Einstein's opinion [3]

Einstein adhered to the more narrow notion of reality of classical physics. From this point of view a description of nature which allows for undetermined single events automatically appeared as "incomplete". He connected it with an aggressive yearning not for the old mechanistic idea of a mass point, but for his geometrisized notion of field of his general theory of relativity. He openly declared his motive: a departure from a more narrow idea of reality of the physics before quantum mechanics appeared to him as a dangerous approach to a point of view in which dream or hallucination could not be clearly distinguished from "reality". On the other hand, the objective character of description of nature by quantum mechanics appeared to us others sufficiently certain by the fact that statistical laws of quantum mechanics describe reproducible phenomena and that the results of observations cannot be influenced by the observer once he has chosen the particular arrangement of his experiment.

2. HOLISTIC ASPECTS OF QUANTUM MECHANICS

In the following we will only be concerned with the *interpretation* of the Schrödinger equation and with quantum mechanics in general. We will not deal with mathematical problems or problems related to calculating and predicting physical experiments and results.The Schrödinger equation, the basic equation of motion of quantum mechanics reads as follows:

$$H\psi = ih\frac{\partial\psi}{\partial t} \tag{1}$$

In the so-called "time-dependent Schrödinger equation" H is the Hamilton-Operator which has a specific form for every concrete problem. ψ is the Schrödinger wave function and h Planck's quantum of action. For simple examples of the Hamilton operator (e.g., the hydrogen-atom or the harmonic oscillator) the Schrödinger equation can be solved exactly.

For more complex problems which cannot be solved in exact terms, Schrödinger developed a method of perturbation calculation to arrive at approximate solutions. These problems are not essential for physics. The problem we are interested in here is the *interpretation* of the wave function ψ in terms of its "description of reality". In classical physics a clear distinction is made between discrete and continuous phenomena. This distinction can no longer be upheld in quantum mechanics. The absolute square of the wave function

$$|\psi(\mathbf{x},t)|^2 = \rho(\mathbf{x},t) \tag{2}$$

can be interpreted - for example in the hydrogen atom - either as charge density of a continuously smeared electron or as the probability of finding a discretely concentrated (point-like) electron at a certain point. Neither of these interpretations is sufficient for describing all phenomena, but both of them are necessary. Depending on the particular issue which the experimental physicist explores with his apparatus, nature answers either within the framework of continuum physics or within that of point-particle physics.

To cite an example: Electrons are scattered with very high energy

$$e^- + e^- \rightarrow e^- + e^- \tag{3}$$

The experimental physicist can determine from the scattering process that the electromagnetic interaction between electrons is as if the charge of the electron were concentrated in a small sphere with a radius smaller than the smallest resolvable length which at present is 10^{-16} cm.[4]

On the other hand, atoms in the lattice of a crystal behave as small spheres with an atomic diameter of about 10^{-8} cm. They thus seem to be "smeared" in a volume the linear dimensions of which are eight times larger than in the high energy scattering process, e.g. (3). Moreover, these discrete and continuous aspects are connected by spontaneous "quantum leaps". A very clear and real physical example is K-capture in an instable nucleus, as for example:

$$_{57}La^{135} + e^- \rightarrow {}_{56}Ba^{135} + \nu_e \tag{4}$$

In a Lantan-atom with an instable nucleus, a 57-fold elementary charge is continuously spread in the atomic shell. We cannot speak of 57 electrons in the shell, since they have no individuality and we would arrive at the wrong physical results if their individuality could not be destroyed mathematically through antisymmetrization of the wave function. From this 57-fold charge shell one complete elementary charge (this corresponds to the discrete character of electric charge!) is spontaneously singled out and concentrated pointlike in one of the protons of the Lantan-nucleus. In the "weak process" both the electron and the proton lose their charge, the proton being transformed into a neutron of the daughter baryum-nucleus and a neutrino emitted.

Schrödinger tried to parodize this interpretation by extending it to macroscopic phenomena. His famous cat[5] has stimulated discussion as Einstein did with his famous paradox[6] which has influenced theoretical ideas up to this very day. It also led up to one of the most beautiful experimental proofs of quantum mechanics.[7]

All of these objections and paradoxa disappear once we loosen the postulate of strict localizability and accept holistic aspects in the microscopic description of nature. Wolfgang Pauli has clearly expressed this[8]:

> In contrast to the fields of classical physics these "probability fields" which have also been called "expectation catalogues" cannot be measured simultaneously at different locations. If a measurement is made at a certain point, it means a transition to a new phenomenon with different initial conditions and thus a new ensemble of possible expectations, therefore a new field everywhere. Hence in atomic physics the phenomena have a property of holism insofar as they cannot be taken apart without changing the whole phenomenon in the process.

And in another context he writes[9]:

> If we try to analyze a phenomenon in separate parts, for example, if we try to follow a certain particle through space and time, we create a new phenomenon because of the uncertain interaction with new apparatuses without reaching the goal of separating the original phenomenon in parts. Thus the phenomenon shows new aspects of indivisibility or wholeness alien to the classical description of nature.

The classical objections against the standard interpretation of quantum mechanics will be dealt with more thoroughly in other papers. Let us examine the further developments of quantum mechanics after Schrödinger.

3. FROM QUANTUM MECHANICS TO QUANTUM FIELD THEORY

Dirac's unification of quantum mechanics and special relativity gave way to a new kind of holism. It became impossible to consider a single particle as an isolated phenomenon. The vacuum became a very complicated physical system, in which each kind of particle is "virtually" present and "real" physical particles interact with this "sea of virtual particles in a vacuum". Without recourse to complicated mathematics we can try to get an idea of the situation: relativistic quantum mechanics predicts that there is an anti-particle for each kind of particle. Pairs of particles and anti-particles can annihilate each other or they can be created. Because of conservation laws, energy and momentum remain conserved in different forms (for example as γ-rays). The creation of particle/anti-particle pairs requires the existence of a sufficient amount of energy (for example, in the form of γ-rays or the kinetic collision energy of other particles). It is typical for the vacuum to have neither energy nor momentum. Nevertheless, particle/anti-particle pairs can be created in so-called "vacuum-fluctuations" because the uncertainty relation of Heisenberg allows deviations from a definite state of energy by a quantity ΔE during a time interval Δt, the product of which is limited by the uncertainty relation

$$\Delta E \,.\, \Delta t < h \qquad (5)$$

If a real physical particle is present in the vacuum, it will interact through its charge through virtual particle/anti-particle pairs. Consequently they will screen its original "bare" charge. This is called the phenomenon of "vacuum-polarization". Experiments have confirmed its existence beyond any doubt. For our considerations it is of particular importance that a particle not only interacts with pairs of its own kind, but with all possible existing particle/anti-particle pairs because all of them are virtually present in the vacuum. Thus the theoretical description of an electron, for example, requires knowledge of the influence of all other particle/anti-particle pairs unless their influence falls below the experimental precision of measurement. A good example is the measurement of the magnetic moment of the electron and its heavy partner, the muon. These measurements have reached a high level of precision.(4) Thus we can say that relativistic quantum mechanics requires knowledge of all particles in order to predict the properties of one given kind of particle.

Finally I would like to mention the so-called gauge theories which in a certain sense represent the quintessence of quantum field theory and relativistic quantum mechanics. Thanks to gauge theories we are able to describe all fundamental interactions of leptons and quarks in one single Lagrange-function (4). Gauge theories show again a new kind of holism: the interactions of elementary particles are predicted from the multiplet structure of free particles. (An elementary description of this idea has been given.[10]) This results in a fascinating possibility: precision measurements of known particles allow a possible prediction whether there are further generations of leptons and quarks, possibly at a level of energy beyond the presently accessible one.

Even Wolfgang Pauli was not yet able to conceive of these new forms of holism. However, a physicist working today finds them quite obvious and uses them in a straight-forward way in his calculations and considerations. It is often so that a very fertile idea yields its best results only in the next or second generation. I am quite sure that Erwin Schrödinger - in spite of his dislike of vapid talk - would have been extremely pleased to see his ideas pursued in the physics of our generation.

NOTES

1 P. Urban, Die Antrittsvorlesung Erwin Schrödingers in Graz über die "Grundidee der Wellenmechanik," 1936.
2 E. Schrödinger, *Mein Leben, meine Weltansicht*, Zsolnay Verlag, Vienna 1985.
3 W. Pauli, *Helv. Phys. Acta. Suppl. IV* (1956), p. 282.
4 D. Haidt and H. Pietschmann, *Electro-weak Interactions*, Springer Verlag, Heidelberg 1988.
5 E. Schrödinger, *Die Naturwissenschaften* 23 (1935), pp. 807, 823, 844.
6 A. Einstein, B. Podolsky, N. Rosen, *Phys. Rev. 47* (1935), p. 777.
7 A. Aspect, J. Dalibard, G. Roger, *Phys. Rev. Lett. 49* (1982), p. 1804.
8 W. Pauli, *Dialectica* 8 (1954), no. 4.
9 W. Pauli, *Dialectica* 11 (1957), no. 1/2.
10 H. Pietschmann, *Phys.Blätter 35* (1979), p. 569.

SCHRÖDINGER'S CAT AND THE INTERPRETATION OF QUANTUM MECHANICS

BERNULF KANITSCHEIDER
Zentrum für Philosophie und Grundlagen der Wissenschaft der Justus-Liebig-Universität
Otto-Behaghel-Straße 10/C II OG., 6300 Gießen, Germany

1. INTRODUCTION

Cats seem to play an important role in physics. The theory of general relativity predicts the existence of collapsed systems which have lost all of their former properties with the exception of mass, charge and angular momentum. These black holes are just a shadow of the former star. Black holes can easily be compared with the cheshire-cat of Alice's adventures in wonderland whose grin hung in the tree even long after the cat had disappeared.[1]

Even in quantum mechanics (QM) cats are often used as a metaphor although some famous physicists are disinclined to do so. Stephan Hawking, e.g., remarked jokingly: "When I hear of Schrödinger's cat I reach for my gun."[2] From a historical point of view Schrödinger's picturesque metaphor is a good starting point for an assessment of his philosophical convictions. What interpretation of QM has been pursued since Schrödinger invented that rather sad story of the cat half alive and half dead?

2. HISTORICAL REMARKS

Already in the beginning of QM's existence there was an extended epistemological debate on the meaning and references of the central terms of that novel physics. The theory came to the fore in two different mathematical formulations that were associated with totally different world views. Heuristics and basic intuitions of wave mechanics and matrix mechanics were of such different nature that it came as quite a surprise that both formulations turned out to be mathematically equivalent. Heisenberg's more empiricist approach was motivated by spectral terms and the particle picture, whereas Schrödinger's operating with classical differential equations that generated discreteness through boundary conditions resulted in a specific spatial function. Schrödinger's preference - the wave picture - cannot be seen as being generated by a mathematical scheme. It is founded on general ontological convictions about "the way things are."

Certain alliances in the historical development seem to be quite natural. In the case of wave mechanics Louis de Broglie's doctoral thesis "Recherche de la théorie des quanta" (1924) was of seminal importance. In this revolutionary piece of research, matter waves were regarded as a fundamental physical reality. Although Paul Langevin and Jean Perrin, under whom de Broglie wrote his thesis, were sceptical as to this new ontology, Einstein realized at once the important progress in this novel idea. Einstein's paper on the quantum theory of a non-atomic ideal gas[3] was the historical link between de Broglie and Schrödinger. Einstein had been interested in a wave theoretical explanation of the Bohr-Sommerfeld quantum rules at a very early date.[4] When Einstein analysed the fluctuations in a quantum gas he was forced to introduce a mathematical

J. Götschl (ed.), Erwin Schrödinger's World View, 41–51.

term the natural meaning of which was a wave field in de Broglie's sense.[5] The same line of development can be observed in the history of science, if we look at Schrödinger's last paper on gas theory before his discovery of wave mechanics.[6]

The decisive step was taken when he seriously considered the ontology of a wave theory of moving particles. Ontological and epistemological assumptions are of much greater importance when a new theory is under construction. By contrast, when new applications and particular solutions are sought, it is possible to make do with minimum empirical interpretation to make the theory testable. On these grounds, it was possible to improve the degree of confirmation of QM without solving the fundamental question as to the reference of the wave function. Born's preliminary interpretation paved the way to a connection of the theory with every empirical fact known at that time. An intermediate instrumentalistic interpretation, though incomplete from a semantical point of view, has some methodological merit in making a theory testable.

Max Jammer was quite right in comparing the early attempts to interpret the formalism with "the mathematical deciphering of a numerical cryptogram in which some of the symbols had been interpreted in accordance with the rules of correspondence of classical physics.[7] A complete semantics of the new theory was at the time seen as the aim of future research. Thus it is understandable that the empirical success of QM grew enormously over the decades whereas the meaning of the crucial terms has remained obscure. It comes as no surprise then that an outstanding theorist like Roger Penrose could utter a statement like the following in 1986: "The theory has, indeed, two powerful bodies of fact in its favour, and only one thing against it. First, in its favour are all the marvellous agreements that the theory has had with every experimental result to date. Second, and to me almost as important, it is a theory of astonishing and profound mathematical beauty. The one thing that can be said against it is that it makes absolutely no sense."[8]

Schrödinger's contribution to formulating an adequate interpretation of his own novel theory exhibits a typically philosophical line. Already in his first paper on wave mechanics[9] he is at pains to explain the quantum rules of Bohr's atomic model in a natural way. Discreteness should be tackled in the same way as the knots of a vibrating string. The boundary conditions of a physical situation generate the particular values of the energy parameter. Regarding the real meaning of QM terms he is rather silent in his first paper, his central concern being to infer the quantum jumps from deeper mathematical principles. The heuristic guiding lines of wave mechanics were modelled on the classical idea of continuity. Not a system of discrete numbers (matrices) but a partial differential equation is the mathematical instrument that rules the physical process.[10] Continuity is without doubt one of the leitmotifs of Schrödinger's life's work as a physicist. In addition, ontological ideas such as unity of nature and epistemological principles such as objectivity are recurring themes of his research. The latter can be seen if we compare Schrödinger's motives of theory construction with those of Louis de Broglie. In this century, quantum theory has encountered two major difficulties: the justification of the stability of the atomic orbit and the duality of light and matter. Duality was at that time the main puzzle. Bohr surmised for some time that the theory of light could be inconsistent on account of the incompatibility of the particle and wave properties. At that time there were intense debates between Bohr and Einstein on a possible experimentum crucis between the classical wave theory of light and Bohr's quantum rules that make essential use of the particle picture. Nobody could imagine at that time how the two aspects could be combined in a coherent way. Einstein defended the construction of a theory combining both pictures which were then, however, to be related in a causal way.

Bohr, though rather disinclined to accept the photon concept in the beginning[11], was so convinced that he subscribed to it, when he realized that the Heisenberg relations avoided inconsistencies in the theory of radiation.

Schrödinger pursued other ontological strategies than Louis de Broglie did. The latter tried to transfer the duality of waves and particles to the properties of matter. All of physical reality should be built symmetrically out of particles and waves. In a sense we can see de Broglie's attempt as being informed by a dualistic ontology. Schrödinger, however, had a monistic ontology in mind.[12] In his opinion particles are not autonomous constituents of nature, but epiphenomena mathematically represented by wave packets. Louis de Broglie, however, saw both waves and particles as being independent elements. He uses the metaphor of a billow carrying a cork: the causal relation of both remains undefined. This lack of causal relatedness might have reinforced Schrödinger's undulatory monism. Causality and a spatiotemporal frame of reference were not the only leifmotifs in Schrödinger's work. Epistemological assumptions were of central importance to him, as well. This becomes evident when we look at his early reference assumption of the wave function. Not ψ itself had meaning; he regarded the term $e\psi\psi^{*}d\tau$ as an electrical density of charge in the volume element $d\tau$. This assumption should allow us to return to a classical electrostatic and magnetostatic model of wave theory, waves being the only physical reality. Discrete energy levels and quantum jumps could then be regarded as intervening variables, mathematical tools with no real reference. In this perspective the quantum postulate had been decoded as a phenomenon of resonance. Nature looks as if there were discontinuous processes in nature, but that phenomenon is only seeming. Ontologically speaking it is an illusion. Only the concept of frequency has a counterpart in reality, energy does not refer to anything tangible. Schrödinger went to great pains to reconstruct all experiments, including those of Frank-Hertz and Compton, in terms of the wave picture.[13]

Regarding the concept of energy he was almost the only physicist to hold that energy is just a macroscopic concept averaging over many elementary processes, whereas this term does not apply to microscopic events.[14]

Schrödinger was quite aware of the fact that he had the burden of proof to explain the apparent particle features and discontinuities appearing in all quarters of physics. He began with the harmonic oscillator and constructed a wave packet of which he could show that this wave group had a kind of inner coherence and might be apt to represent a particle.[15] He hoped that the wave mechanical picture could also be applied to the hydrogen atom. The problem was, of course, stability. Does the wave entity which corresponds to the particle remain within a bounded spatial region without dispersing with time? Even a member of the older generation of physicists, namely Lorentz, doubted this.[16]

It was Werner Heisenberg who showed that Schrödinger's approach to constructing wave packets for any potential is limited. Schrödinger's attempt to attain classical objectivity and realistic description encountered further obstacles. Physical space has three dimensions, and there is no indication that higher spatial dimensions are needed for any known phenomenon (this was, of course, before the advent of supergravity and superstring theory.) If wave mechanics is applied to n-particle systems, representational space acquires 3n dimensions. The question therefore remains unanswered as to how to relate these 3n dimensional wave processes to physical 3-space.[17]

In opposition to classical dynamics $|\psi\rangle$ is a complex variable and there must be some physical sense of this property.[18]
The last obstacle for a simple realisitic objectivity of ψ is its dependence on the pertaining observable. The meaning of ψ is quite different, depending on which operator is chosen in representing ψ.

Schrödinger disagreed strongly with Born's probabilistic interpretation of $|\psi\psi^*|$, because Born was more or less guided by the particle picture.[19] $|\psi\psi^*|\, d\tau$ means that a particle in the state ψ may be found in an element of the volume of configuration space $d\tau$ with a certain probability. This is, of course, an epistemic turn away from an ontological interpretation of $|\psi\rangle$, because QM provides an answer to the question not of the state of a system, but of the probability of finding the system in a certain state after the collision. The causal spatiotemporal description has been shifted from the ontic to the epistemic level of probabilities.[20]
This means that e.g., a singular level transition of energy within the atom is no longer causally determined even if it possesses a high a priori probability. The transition itself, the jump, is a non-analysable elementary phenomenon.[21] Schrödinger felt quite uneasy about the fact that a quality of man, i.e., his knowledge, appeared amidst physical semantics[22], although he realized that Born's interpretation overcame various difficulties of a direct realistic interpretation of ψ. The above-mentioned obstacles with regard to a tangible meaning of ψ disappear, if an epistemic turn is assumed, but Schrödinger refused to weaken the sense of ψ as the key term of QM. Later on it became clear that Schrödinger's intuition was not completely off track because Born's interpretation, very appropriate for scattering problems[23], cannot be so easily applied to situations where only one particle is in the experimental set up and the wave function shows self- interference[24]. For this reason Heisenberg was inclined to associate ψ with a dispositional tendency, a propensity of matter exhibiting certain properties. Schrödinger, however, could not cope with the Kirchhoff-Mach-descriptivism[25] that was the background philosophy of the two proponents of the Copenhagen interpretation at that time.

3. EPR AND THE CAT

Metaphysical ideas that act as catalysts for physical theories are only vicariously testable, when the theories in question are well-confirmed. Metaphysical background convictions can prove to be very obstinate. The feedback from their failure to form a successful scaffolding for empirical theories shows a very long time lag. The tenacious efficacy of philosophical beliefs and the persuasive power of the background assumptions can be easily recognized in the conceptual quarrel of the EPR-argument.[26]

Schrödinger was the first to call it a paradox. Philosophers of science speak of a paradox when correct deductive reasoning leads from plausible assumptions to untrustworthy consequences. The EPR-paper triggered Schrödinger once again to deal with the foundations of QM. In his paper "Discussions of Probability Relations between Separated Systems"[27], however, he had a different goal than Einstein. In collaboration with his co-workers he tried to prove the incompleteness of QM. Both agreed, however, that a realistic model had to be found for QM[28], but Einstein never took QM as a fundamental, but only as a preliminary theory[29] to be replaced later on by a unified field theory, the central concept of which should be "overdetermination."[30]

Schrödinger acknowledged the fundamental character of QM and focussed on certain traits of this theory, namely the interlocking parts. In QM partial systems of a whole are concatenated in a characteristic non-classical way. The interrelationship of the parts of larger quantum mechanical systems described by a total wave function does not allow the factorization to its partial wave functions. Here we encounter the first clear account of the obscure commonsense wisdom: "The whole is more than the total of its parts." As Hans Primas has put it: "Quantum mechanics is the first mathematically formulated scientific theory, where, in a non-trivial way, the whole is more than the sum of its parts."[31]

On the other hand, there is something strange about this quantum mechanical situation, because an observer, through measurement of a partial system, can steer the other subsystem to a certain state without physical interaction with another remote part. Schrödinger had a presentiment of the runaway situation hidden in the disentanglement that later on was called the v. Neumann catastrophe. "Disentanglement is of sinister importance ... threatening us thereby with at least a regressus in infinitum..."[32]

Within J. v. Neumann's thorough analysis of quantum mechanics of measurement (QMM) there are two kinds of dynamics of the quantum mechanical state vector. The borderline between both (the so-called cut) can be varied at will. The collapse of the state vector may occur within the experiment, in the apparatus or in the consciousness of the observer.[33]

In order to demonstrate the paradoxical situation brought on by a macroscopical amplification of a microscopic measurement, Schrödinger came up with his famous example of the cat.[34] Basically, it is an application of the superposition principle of QM. After one hour the whole sealed steel chamber with the cat in it is described by a wave function that is the superposition of two states

$$\psi = (\psi_L + \psi_D)$$

In our macroscopic world we never encounter such hazy situations in which two incompatible properties, a live cat ψ_L and dead cat ψ_D are inextricably entangled.[35] Although Schrödinger was convinced that there must be some deficiency in the basic structure of QM he could not point to the exact cause. He surmised that the reason could be the concept of time in QM, quantum mechanical time being a classical parameter and not a unitary operator. Only considerable time later was it realized how classical variables could be handled in general QM.

4. THE FURTHER FATE OF THE CAT

Schrödinger's cat caused quite a stir in the philosophical community. There were a number of commentaries, some affirmative reactions like Einstein's, some hesitant and some disinclined ones, like Hawking's, as we saw in the introduction. It is essential to note that the pure state of superposition of ψ_L and ψ_D is fundamentally different from a mixture of both states of which the observer is ignorant. A superposition means that before the measurement both possibilities coexist, with only the stochastic process of measurement yielding $\psi = \psi_L$ or $\psi = \psi_D$. There is never more than one possibility where the collapse of the wave function could occur. The human observer may, for example, trigger that event when he opens the steel box. Hilary Putnam pointed to the possibility that the cat could be regarded as quantum mechanical observer.[36] Moreover, the irreversible amplification of the radioactive decay process may cause the stochastic transformation from the superposition to the mixture. At any rate, to say that macroscopic variables cannot be in a state of superposition reflects a dogmatic attitude that limits the validity of QM without necessity. Recourse to common sense is rather out of place in the quantum mechanical debate. The idea of amplification is likely to prove more successful. Reduction occurs according to John Wheeler "when the quantum mechanical phenomenon has been brought to a close by an irreversible act of amplification."[37] The thermodynamical properties of the apparatus cause the transition from the superposition to the mixture. J. v. Neumann introduced an additional measurement postulate, not contained in the axiomatics of QM, that the wave function turns into the proper state of the operator upon observation. If this postulate of QMM is neglected and the Schrödinger equation applied to macroscopic systems, then the dynamics of the combined system of cat and observer develops into a still larger superposition

$$\psi_{\text{cat+observer}} = (\psi_D \times \psi_{\text{observer sees a dead cat}}) + (\psi_L \times \psi_{\text{observer sees a live cat}}) \quad (1)$$

The result of quantum dynamics seems to be preposterous, because the observer himself has to be conscious of both contradicting visual impressions. Even after the observation the case remains undecided. That was Eugene P. Wigner's starting point for his mentalistic interpretation of QM.[38] Since it is impossible for a human mind to be "in a state of suspended animation", the decision between the two possibilities

$$\psi = \psi_D \times \psi_{\text{observer sees a dead cat}}$$
or
$$\psi = \psi_L \times \psi_{\text{observer sees a live cat}} \quad (2)$$

must have been obtained at the moment the observer opens the steel box. Superposition (1) is the correct description only for a material measurement apparatus, an intelligent organism proceeds immediately to the mixture (2).

Wigner's solution of Schrödinger's cat paradox has far-reaching consequences for our understanding of nature. First of all, the linear equation of QM cannot be extended to organic and mental systems. Wigner's antinaturalistic approach is also incompatible with a neurological theory of mental events. Within such a monistic account the sensational impressions of the observer are identified as a complex pattern of firing neurons. If one day that activity of the neuron could be described in quantum mechanical terms, it would only lead to a gigantic superposition of neurophysiological states. Only if mental events are couched in a dualistic language would the act of sensation effect the collapse. Since the dualistic approach to the mind-body problem is in a rather problematic state - pace Popper/Eccles - and since there are many more adherents of a materialistic monism, the mentalistic solution to the central problem of QMM is not very promising. Beyond that there are some snags in the antinaturalistic approach. Where in the rather long evolution of the brain should we draw the line between a physical registration system that cannot reduce wave packets and an organic system with the emergent property of being able to do so? Who has the legitimation to reduce wave packets? There is also the familiar difficulty due to special relativity theory pointed out by Abner Shimony[39] given when there are two observers looking at the cat. Both can take snapshots of their observations and carry them to spacelike parts of the universe. According to special relativity theory it then depends on the reference system which observer looks first and reduces the superposition. It is cogent that both sequences of observation would lead to the same experimental result.

The most speculative solution of Schrödinger's cat paradox has been proposed by John Barrow and Frank Tipler in 1986.[40] Already Hugh Everett pointed to the unpleasant runaway situation which occurs if many observers are connected in series. Under ordinary physical conditions, in which the mental abilities of an observer are not invoked, the measurement process cannot be completed. Barrow and Tipler now assume now that there is a chain of observers that extends into the future until at a certain point it meets observer O_j who belongs to another chain. O_j is connected with O_{j+1} in the distant future and so on. Pursuing this idea, we can picture a network of observers of all intelligent living beings in the universe. To install a final link of the many chains of this hierarchy of observers, Barrow and Tipler postulate the so-called *ultimate observer*. The ultimate observer has a special status. He can afford the final observation because he is located at the final singularity in a closed universe and in the time-like infinity in an open universe. On account of the global structure of general relativistic space-time there can be no further observation, singularities and timelike infinity being boundaries of space-time. No additional term in the series of observations is possible.

Barrow and Tipler specify the whole procedure mathematically in the comprehensive formula

$$\psi = \prod_{n} \sum_{i,k} \psi_{ik} O_{ink} \tag{3}$$

i = index for the possible outcome of kth experiment
n = index for the sequence of observers that look at the kth experiment

The ultimate observer reduces the activity of every earlier observer, a unique result of the index i being the outcome. He is not part of the universe, because he is outside the space-time manifold. Nobody can tell in advance which of the indices (i) in the formula (3) will be realized before the universe has reached its final state. Only then, the quantum mechanical actualization of the universe will be definite. Objectification of our whole existence will happen in finite time only for the closed (compact) universe. Barrow's and Tipler's solution is extremely idealistic, consciousness being constitutive for the existence of reality. Before the ultimate observer does his duty there is no quantum mechanical actualization of the whole system. Above that a supernaturalistic interpretation of the ultimate observer seems obstruse. Nobody will take this proposal seriously as a solution to the measurement problem. However, it illustrates very well the inapplicability of the v. Neumann approach within a physical reality conceived in naturalistic terms.

5. THE CAT THAT WALKED BY HIMSELF

How can one proceed given such problematic conditions? The decisive question seems to me to be whether we should see the transition from superposition to single states as an epistemic or an ontic step, a process happening in the mind of the observer or in physical reality.

Hilary Putnam opted for the epistemic alternative exploiting Schrödinger's cat as a reason for his internal realism.[41] The situation suggests a certain kind of epistemic relativity. Either we accept only the external observer view as the pertinent framework to tackle with the situation or we regard the cat himself as a quantum mechanical observer. According to Putnam, this is a case of relative truth. We can remain realists on internal issues with regard to certain properties of the cat system, but the nature of these properties is dependent on the framework, i.e., the observer. There can hardly be any doubt that the original intuition and the core meaning of epistemic realism is lost if they become relative in that way. Neither Einstein nor Schrödinger would have been content with Hilary Putnam's slogan: "There are real entities but what they are is up to the observer."[42]

In this context it is a common strategy to point to special relativity theory. Even Bohr himself[43] defended the relativitiy of the quantum mechanical state description by means of an analogy argument from relativity theory. Nevertheless, this justification is unsound, because both kinds of relativity are quite different. In the Copenhagen view quantum mechanical states are related to experimental apparatus, whereas in relativity theory certain variables depend on the reference system. Reference systems are connected by mathematical transformations that enable us to compute the values of the dynamical variables in every other system. Beyond that, there are invariant magnitudes that do not depend on the reference system one chooses. In QM, however, there is not such a continuous transition from one experimental set up to the complementary one. Besides this, there are no absolute magnitudes like the invariant line element ds in relativity theory. There is no insight as to what happens causally if we change from one experimental set up to the next. In the Bohr-Putnam internal approach we encounter a relativity of the concept of property, but it is not really anchored in the mathematical structure of the formalism. The Hilbert space reflects the quantum mechanical states quite impartially without any complementary fragmentation. The mathematical structure does not contain any indication of a relative description that would force us to introduce the rather weak internal realism. Therefore, it is quite natural if we take the formalism of QM as a starting point for finding a solution to the observer-system interaction which is in keeping with a stronger epistemological commitment. Where do we find the second alternative that deals with reduction as a real ontic process?

6. REDUCTION OF AN EFFECT OF QUANTUM GRAVITY

Schrödinger would not have agreed with that lame sort of realism which is strongly advocated in recent analytical philosophy.[44] It is interesting to note that not even mathematical physicists who are deeply engaged in solving this problem from the perspective of physics adhere to a stronger non-relativistic realism. Worth mentioning in this context are the attempts to regard the reduction of the state vector as an effect of quantum gravitation. Roger Penrose, who together with S. W. Hawking proved the singularity theorem of general relativity, proposed that the collapse of the wave function may be caused by gravitational interaction.

What is decisive about this approach is that ψ is taken seriously in terms of physics and that its collapse is seen as a real dynamic process. Penrose's demand for physical objectivity of the state vector and dynamic reality of the reduction cannot be fulfilled within the mathematical framework of conventional QM because, as we have noticed with the example of Schrödinger's cat, that measurement leads to an unlimited runaway process. Penrose takes the superposition of the dead and live cat as an indication that physical reality is not completely described by the evolution of the state vector according to Schrödinger's equation. At the level of the cat, something is missing which is not taken into account by the state vector. Schrödinger's equation becomes an unprecise description with the character of an approximation. Which feature of reality managed to evade description through Schrödinger's development? Many authors have agreed that a measurement is always linked with an irreversible act of amplification, i.e., a thermodynamic process that leads to an increase in entropy. Can we possibly objectify the thermodynamic component in the measurement process? There is one special domain of physics where we encounter a well-defined objective concept of entropy, namely within the thermodynamics of a black hole.

The entropy of a black hole

$$S_{bh} = \frac{kAc^3}{4hG}$$

can be formulated in terms of a space-time concept, A being the surface of the black hole on the event horizon. The measure of entropy is a totally objective feature of space-time geometry. The entropy of the gravitational field in general is, according to Penrose, the pertinent agent that causes the reduction of the state vector. Although until now no mathematical object is known to yield the gravitational entropy in general, the Weyl curvature $C^h{}_{ijk}$ might play a central part in it. The tensor $C^h{}_{ijk}$ is defined in the following way

$$C^h{}_{ijk} = R^h{}_{ijk} + 1/2\,(\delta^h{}_j R_{ik} - \delta^h{}_k R_{ij} + g_{ik} R^h{}_j - g_{ij} R^h{}_k) + R/6\,(\delta^h{}_k g_{ij} - \delta^h{}_j g_{ik})$$

$C^h{}_{ijk}$ = Weyl-tensor
$R^h{}_{ijk}$ = Riemann- tensor
R_{ik} = Ricci- tensor
R = Riemann - scalar

Quantum mechanical state reduction is induced by a growth of gravitational entropy. More precisely, "reduction will occur, when the decrease in entropy involved in the reduction effect is at least compensated by a corresponding increase in gravitational entropy."[45] This additional process presumably cannot be covered by ordinary linear QM and its unitary dynamics. Moreover, as must be surmised, if curved space-time enters the quantum mechanical picture, superposition of states with different geometries will not likely occur. Instead it is reasonable to assume that nature tends to realize just one peculiar geometry. Penrose thinks of some nonlinear instability as being responsible for the transition.

It is perhaps interesting to note that other research groups have searched for a gravitational solution to the quantum mechanical measurement enigma, as well. F. Karolyhazy, A. Frenkel and B. Lukacs have expressed the basic idea[46] that at a certain level of complexity stochastic elements intrude the dynamical evolution of singular systems. Schrödiger's evolution of complex systems tends towards building superpositions, whereas stochastic tendencies try to reduce these superpositions to mixtures. The crucial point is to know in which domain of complexity the stochastic influence becomes dominant.

Surprisingly, already in 1935 Schrödinger seemed to have a foreboding of this idea[47] when he claimed that the discontinuous jump within the measurement is the major obstacle for scientific realism. It is for this reason that we cannot regard ψ as the model of reference mapping a real thing. Observation, he claimed, is a natural process itself and should not interrupt the course of development of nature in an unintelligible way.

Recent attempts to combine two kinds of evolution, causal and stochastic, in the quantum mechanical process can be seen as a continuation of Schrödinger's original intuition. From a philosophical point of view it is decisive that the stochastic reduction of the extended wave function to one of its minor parts has been carried out by nature itself and not by the magical abilities of the observer. This does not mean to belittle the merits of QM. Even if the theory is telling the truth, it need not be the whole truth. Introducing gravitational interaction in the quantum mechanical description could restore objectivity and any premature break with realism would be unnecessary.

We are well advised to take the early warnings of Einstein and Schrödinger seriously in order not to prematurely give up an essential claim to knowledge without being compelled to do so.

NOTES

1 Lewis Carrol, *Alice's Adventures in* Wonderland, New York 1960.
2 J. Barrow; F. Tipler, *The Anthropic Cosmological Principle*, Cambridge 1986, p. 458.
3 A. Einstein, "Quantentheorie des einatomigen idealen Gases," 2. Abh. Sitzber. Preuß. Akad. Wiss. Berlin 1925, pp. 3-14.
4 cf. his letter to H. Lorentz from Dec.16, 1924.
5 K. Przibram (ed.), *Briefe zur Wellenmechanik*, Vienna 1963, p. 24.
6 E. Schrödinger, "Zur Einsteinschen Gastheorie," *Physikalische Zeitschrift 27* (1926), pp. 95-101.
7 M. Jammer, *The Philosophy of Quantum Mechanics*, New York 1974, p. 23.
8 R. Penrose, "Gravity and State Vector Reduction," in: R. Penrose; C. J. Isham (eds.), *Quantum Concepts in Space and Time*, Oxford 1986, p. 129.
9 E. Schrödinger, "Quantisierung als Eigenwertproblem," *Annalen der Physik 79* (1926), pp. 361-376.
10 E. Schrödinger, "Über das Verhältnis der Heisenberg-Born-Jordanschen Quantenmechanik zu der meinen," *Ann. Phys.* 9 (1926), pp. 734-756.
11 N. Bohr, H. A. Kramers und J. C. Slater, "Über die Quantentheorie der Strahlung," *Zeitschrift für Physik 24* (1924), pp. 69-87.
12 E. Schrödinger, "Zur Einsteinschen Gastheorie," *Physikalische Zeitschrift 27* (1926), pp. 95-101.
13 E. Schrödinger, "Energieaustausch nach der Wellenmechanik," *Ann. Phys. 83* (1927), pp. 956-968.
14 E. Schrödinger, "Might perhaps energy be a merely statistical concept?" *Nuovo Cimento 9* (1958), pp. 162-170.
15 E. Schrödinger, "Der stetige Übergang von der Mikro- zur Makromechanik," *Die Naturwissenschaften 14* (1926) pp. 664-666.
16 Cf. the letter from Lorentz to Schrödinger from June 19, 1926, in: K. Przibram, *Briefe zur Wellenmechanik*, loc. cit. p. 61.
17 M. Jammer, *The Philosophy of Quantum Mechanics*, New York 1974, p. 33.
18 Only later did it become quite clear how complex phases relate to phenomena of interference.
19 M. Born, "Zur Quantenmechanik der Stoßvorgänge," *Zeitschr. f. Physik 37* (1926), pp. 863-867.
20 M. Born, "Zur Quantenmechanik der Stoßvorgänge", *Zeitschr. f. Physik 38* (1926), pp. 803-827.
21 M. Born, "Das Adiabaten-Prinzip in der Quantenmechanik", *Zeitschr. f. Physik 40* (1926), pp. 167-192.
22 E. Schrödinger, *Briefe zur Wellenmechanik*, ed. Karl Przibram, Vienna 1963, p. 17.
23 For a more profound analysis of Born's interpretation cf.A. Bartels; M. Stöckler, "Die Bedeutung von Wahrscheinlichkeitsaussagen und die Interpretation der Quantenmechanik," in: W. Kuhn (ed.), *Vorträge der Deutschen Physikalischen Gesellschaft Frühjahrstagung, Berlin 1987*, DPG FA Didaktik der Physik.
24 cf. F. Selleri, *Die Debatte um die Quantentheorie*, Braunschweig 1983, p. 77.
25 F. Selleri, *Die Debatte um die Quantentheorie*, loc.cit., p. 115.
K. Przibram (ed.), *Briefe zur Wellenmechanik*, loc. cit., p. 18.
26 A. Einstein; B. Podolsky; N. Rosen, "Can Quantum-Mechanical Description of Physical Reality be Considered Complete?" *Phys. Rev. 47* (1935), pp. 777-780.

27 E. Schrödinger, "Discussions of Probability Relations between Separated Systems," *Proceedings of the Cambridge Philosophical Society 31* (1935) pp. 555-562.

28 E. Schrödinger, "Der erkenntnistheoretische Wert physikalischer Modellvorstellungen," *Jahresbericht des Physik.Vereins zu Frankfurt 1928/1929*, pp. 44-51.

29 The reason for this was that QM did not agree with Einstein's "themata" of theory construction (cf. B. Kanitscheider, *Das Weltbild Albert Einsteins*. Seine Physik und seine Philosophie, Munich 1988.)

30 Cf. A. Einstein, "Bietet die Feldtheorie Möglichkeiten für die Lösung des Quantenproblems?," *Sitz. Ber. der Preuß. Akad. Wiss. Phys.-math. Klasse*, pp. 359-364.

31 H. Primas, *Chemistry, Quantum Mechanics and Reductionism*, Berlin 1981, p. 143.

32 E. Schrödinger, "Discussions of Probability Relations between Separated Systems," *Proceedings of the Cambridge Philosophical Society 31* (1935), pp. 555-563.

33 F. London and E. Bauer, *La Theorie de l'Observation en Mécanique Quantique*, Paris 1939.

34 Within ordinary quantum mechanics the superposition principle is applied without restrictions; restrictions are only possibly by way of superselection rules, in this case some operators cannot be regarded as quantum mechanical observables.

35 E. Schrödinger, "Spezielle Relativitätstheorie und Quantenmechanik," *Berliner Berichte* 1931, pp. 238-248.

36 H. Putnam, "A Philosopher looks at Quantum Mechanics", in: R. G. Colodny (ed.), *Beyond the Edge of Certainty*, Prentice Hall Englewood Cliffs, N. J. 1965, pp. 75-101.

37 J. A. Wheeler, "Law without Law", in: J. A. Wheeler, W. H. Zurek (eds.), *Quantum Theory and Measurement*, Princeton 1983, p. 182.

38 E. P. Wigner, "Remarks on the Mind-Body Question," in: *Symmetries and Reflections*, Bloomington & London 1967, pp. 171-184.

39 A. Shimony, "The Role of the Observer in Quantum Theory", *A. Journ. Phys. 31* (1963) pp. 755-773.

40 J. Barrow, F. Tipler, *The Anthropic Cosmological Principle*, Oxford 1986, p. 471.

41 H. Putnam, "Quantum Mechanics and the Observer," *Erkenntnis 16* (1981), pp. 194-218.

42 H. Putnam, "Quantum Mechanics and the Observer," loc. cit., p. 218.

43 N. Bohr, "Kausalität und Komplementarität," *Erkenntnis 6* (1906), pp. 293-303.

44 Cf."Schrödinger's Use of Realism" in: W. Heisenberg, M. Born, E. Schrödinger, P. Auger, *On Modern Physics*, Opotter N.Y. 1959, p. 38.

45 R. Penrose, *Gravity and State Vector Reduction*, loc. cit., p. 141.

46 F. Karolyhazy, A. Frenkel and B. Lukacs, "On the Possible Role of Gravity in the Reduction of the Wave Function," in: P. Penrose and C. I. Isham (eds.), *Quantum Concepts in Space and Time*, Oxford 1986, pp. 109-122.

47 E. Schrödinger, "Die gegenwärtige Situation in der Quantenmechanik," *Die Naturwissenschaften 23* (1935), pp. 807-849, par. 7.

THE SCHRÖDINGER FUNCTION OF MACROSCOPIC SYSTEMS

KURT BAUMANN
Institut für Theoretische Physik der Karl-Franzens-Universität
Universitätsplatz 5, 8010 Graz, Austria

My contribution will focus on the following question: Has the discovery of macroscopic quantum phenonomena brought about a change in our understanding of nature?

In the early days of quantum mechanics, the new mechanics was called atomic mechanics, since it was assumed that Newton's classical mechanics had to be replaced by a new kind of mechanics in the atomic domain. Conversely, physicists were convinced that beyond the atomic domain quantum mechanics and classical mechanics are virtually indistinguishable. Niels Bohr even claimed that the classical description holds strictly for massive bodies because classical physics is logically prior to quantum mechanics.

A new situation arose in 1957, when Bardeen, Cooper, and Schrieffer successfully explained superconductivity in terms of quantum mechanics. They used a single three-dimensional matter wave to describe the electrons in a superconductor. This wave is characterized by wave length, amplitude and phase. Amplitude and phase of a matter wave are connected by an uncertainty relation: The more precisely the amplitude (or equivalently, the number of particles) is known, the less well-defined the phase is and vice versa. The matter wave (or Schrödinger function) of Bardeen, Cooper and Schrieffer describes a quintillion (10^{18}) electrons or more, and thus represents a macroscopic phenomenon. In defining its properties measuring errors are inevitable, which by far exceed the error limits imposed by quantum mechanics.

The existence, in the superconductor, of a matter wave with macroscopic amplitude seems to show that even beyond the atomic domain quantum mechanics may have consequences that have nothing to do with classical mechanics. In 1980, Anthony J. Leggett [1] questioned this conclusion. He pointed out that in a superconductor 10^{18}, matter waves with microscopic amplitude cooperate to produce a macroscopic wave. Each microscopic wave describes an electron pair. Consequently, superconductivity implies only that the dynamics of a single electron pair is described correctly by the Schrödinger equation.

The existence of macroscopic Schrödinger functions, Leggett argued, could be demonstrated by detecting the quantum behavior of macroscopic oscillators. An experiment of this kind may be carried out in the following way: an electrical current in a superconductor does not lose energy, i.e., there is no resistance. If the superconductor is a closed loop the current is, in addition, quantized - its magnitude belongs to a discrete sequence of values. This is again due to the properties of the individual electron pairs: there is a discrete sequence of possible microscopic wave functions. This can be compared with the electron of the hydrogen atom and its discrete sequence of stationary states.

The crucial point is that the current is not exactly equal to one of the quantized values. Rather, it is attracted by these values and performs oscillations centered on them. These oscillations can be studied for determining macroscopic quantum effects.

J. Götschl (ed.), Erwin Schrödinger's World View, 53–56.

A highly interesting quantum phenomenon to be expected is the tunnel effect. It is based on the fact that quantum mechanics permits transient violations of energy conservation. For example, the current sometimes deviates exceptionally far from the quantized value, thereby violating energy conservation. It can approach the neighboring quantum value and then continue to oscillate about the new center. This phenomenon was observed, and the frequency of the current jumps agrees with the predictions of quantum theory.[2] This means that even massive bodies obey quantum mechanics rather than classical mechanics.

But it also means that the very non-quantum-mechanical notions held by most physicists concerning massive objects are no longer tenable. Leggett and Garg[3] have summarized these notions as follows:

1) Macroscopic realism: a macroscopic system with two or more macroscopically distinct states at its disposal will at all times be in one or the other of these states.
2) Noninvasive measurability at the macroscopic level: it is possible, in principle, to determine the state of the system with arbitrarily small perturbation of its subsequent dynamics. (In quantum mechanics, the state of a system is in general carried over into a different state if the system is subjected to a measurement.)

The macroscopic tunnel effect disproves assumption (1). In order to see this we can ask how the current penetrates the energy barrier. A weak extension of the matter wave enters the forbidden zone. As the wave proceeds, its intensity decreases rapidly so that only a very faint wave reaches the next allowed domain. The main part of the wave is still on the other side of the forbidden zone, i.e., in the vicinity of the original quantized current value. A Schrödinger function of this kind describes a state of the superconductor in which a macroscopic property (the electrical current) has no definite value. Roughly speaking, the current is a mixture of two values. The state is to be compared with the state of Schrödinger's cat. It disproves assumption (1). Later it will be shown that assumption (2) is violated as well.

The invalidity of the two assumptions deprives macroscopic properties of their distinguished position in quantum mechanics. Especially, the quantum mechanical theory of measurement must be reconsidered. At which stage of a measuring process does an unequivocal measuring result come into existence? If assumption (1) were correct the result would arise as soon as a macroscopic quantity becomes involved in the measuring process. In the meantime we know that occasionally macroscopic variables are subjected to a quantum measurement. Consequently, the measuring result must appear at a different stage of the measurement.

Every quantum measurement essentially involves some sort of irreversible process. For some time it has been assumed that this process is accompanied by the appearance of a definite measuring result. This is well confirmed by the macroscopic tunnel effect: massive objects are always dissipative because friction is unavoidable, unless the object is in its state of lowest energy and hence cannot lose energy. In the initial state of the macroscopic tunnel effect we have exactly this situation: the electric current is smeared out with minimal uncertainty around a quantized value. Only after tunneling through the forbidden zone is emission of energy quanta possible again. After a very short time the lost energy becomes large enough to be registered without any quantum measurement. At this stage, the measurement is carried out and a current jump can be observed.

The tunneling decreases if dissipation increases. This phenomenon disproves assumption (2): the energy quanta are a kind of observation means, and the observation changes the behavior of the current.

Dissipation is not the only irreversible process capable of producing a measuring result. The quantum state of an individual atom, for example, may be observed by light scattering. It is possible, by selecting the right wave length, to produce an extremely large scattering intensity in any selected quantum state. The phenomenon is called resonance fluorescence. The number of light quanta scattered per second amounts to up to a billion. Although this is still tiny compared to the frequency of the light wave (600 trillion cycles per second) it is enormous in comparison to a normal scattering rate of one light quantum per second.

The resonance radiation emitted by a single atom is distinguishable from an equally intense radiation emitted by many atoms. In the former case bunches of light quanta are strongly suppressed while in the latter case they are not. Resonance fluorescence is an irreversible process, i.e., it is impossible to transform the spherical light wave into a plane wave with no "antibunching." As a consequence it is impossible to devise an experiment which demonstrates that the state of the electron together with its radiation field is a superposition of different quantum states. Therefore this superposition is indistinguishable from a "mixture" of quantum states.

Each concomitant state of the superposition is an energy state of the atom surrounded by the appropriate radiation field. According to the quantum theory of measurement, only combined measurements of the atom and the scattered light are suitable for demonstrating that there is a superposition rather than a mixture. One can never find interference effects between the radiation fields belonging to different atomic states by observing only the radiation fields.[4] As a consequence of the irreversibility of the scattering process, even a combined measurement does not yield any interference effects.

If we maintain that the measuring result becomes "actual" through an irreversible process we can conclude that one of the scattered fields is the relevant one. It is reasonable to classify these fields as microscopic because they are the contributions of a single atom to the scattering of light by a gas. Taken together, these two statements signify the decoupling of the notions of "actual" (or "real") and "macroscopic" in another way: a microscopic radiation field may be an element of reality whereas a macroscopic current may be the object of a quantum measurement.

Considering that microscopic fluorescence radiation may play the role of a "pointer," there is no longer any point in stressing the macroscopic nature of the measuring device. Of course, an observation of the measuring device is necessary in case the device is microscopic. Thus, for the registration of the fluorescent light we need a photon counter. This registration, however, is not a quantum measurement with the fluorescent light playing the role of the system to be observed. Rather, it is an ordinary observation in the classical sense, i.e., the perturbation of the system by the observation may be kept arbitrarily small if care is taken to ensure that only a small fraction of the scattered light is absorbed.

NOTES

1 A. J. Leggett, "Progress of Theoretical Physics," *Suppl. 69* (1980) 80.

2 R. F. Voss and R. A. Webb, *Phys.Rev.Lett.47* (1981) 265, and S. Washburn, R. A. Webb, R. F. Voss, and S. M. Faris, *Phys. Rev. Lett. 54* (1985) 2712 confirmed the macroscopic tunnel effect by observing the time behavior of the phase difference across a Josephson junction. This phase difference is a macroscopic quantity.

3 A. J. Leggett and A. Garg, *Phys. Rev. Lett. 54* (1985) 857.

4 The properties of this radiation were studied by J. Javanainen, *Phys. Rev.* A33 (1986) 2121, and by A. Schenzle, R. G. DeVoe, and R. G. Brewer, *Phys. Rev.* A33 (1986) 2127. These authors did not make use of the fact that the wave function of the combined system (atom plus radiation field) is a mixture with respect to field observables alone. Rather, they verified this fact by direct computation.

PART II

LIFE SCIENCES AND THE THEORIES OF SELF IN ERWIN SCHRÖDINGER'S PHILOSOPHICAL SYSTEM

SCHRÖDINGER'S NEGENTROPY CONCEPT AND BIOLOGY

RUPERT RIEDL
Institut für Zoologie der Universität Wien
Althanstraße 14, 1090 Wien, Austria

To ask a biologist to contribute to a discussion on a physicist is a bold step. Don't we run the risk of hopelessly confusing the issue? We can only hope that the topic at hand has remained unclarified precisely because too few biologists and physicists have talked to one another. Be that as it may: culture can be understood as a dialogue between disciplines. And this is what I wish to focus on. With all due modesty, I will base my talk on the interpretation of a footnote Schrödinger added to one of his later works.

"My remarks on negative entropy" Schrödinger observed, "have met with doubt and opposition from physicist colleagues..." But, as he continued, "entropy with a negative sign is not something I invented. In actual fact it is what Boltzmann's original argument revolved on."

The cited lines are taken from a more detailed footnote - or note - that Erwin Schrödinger added to the second edition of his lecture series *What is Life?* from 1943. It refers to chapter 6 ("Order, Disorder and Entropy") of his book.

Schrödinger asked a fundamental question: what is negative entropy? Why the opposition? Apparently because here our comprehension of the foundations of this world is also called into question. It is fundamental for our understanding of life and remains essentially unclarified to this day.

The story I want to unravel begins in the first week of September 1906. The 62-year old professor of physics at the University of Vienna, Ludwig Boltzmann, had just committed suicide and the 19-year old high school graduate Erwin Schrödinger had just entered the very same university to become a phyisicist. Perhaps it was written in the stars. The stage had been set. Never again was Schrödinger able to elude Boltzmann's influence. (Kerber et al., 1987)

WHAT IS NEGATIVE ENTROPY?

The notion of entropy is generally said to date back to Clausius (1850). Conceived as the relative measure of that part of heat energy which cannot be converted into mechanical work, it is lost and results in the irreversibility of a process (Greek: entrepein - turn toward.) This interpretation - stemming from thermodynamics - is not very descriptive since it mainly refers to the efficiency of heat engines.

Only sixteen years later, 1866, Boltzmann developed an entirely new approach, one rooted in statistical mechanics. It was more descriptive in that it allows us to imagine the distribution of a number of gas molecules darting about in a chamber. For example, we can calculate the degree of improbability that, by chance, all these particles would be in the right half of the chamber at the same time.

The merit of this approach is not only in its descriptive nature. Energy, work, and irreversibility remain integral components here. At the same time it provides insight into the relation of at-

J. Götschl (ed.), Erwin Schrödinger's World View, 59–69.

tainable and unattainable knowledge or predictability about the states of a system and expected changes in its state.

Maxwell's demon can be used to illustrate this relation. The demon can be imagined to be sitting inside a two-chambered container. Every time a molecule darts from the right to the left chamber, he opens the door between the two. The door is shut when a molecule heads in the other direction. Paradoxically, he could utilize the resulting pressure difference to power a machine. The machine is then capable of doing an amount of work equivalent to the demon's knowledge of the molecules' movement - information to which we ourselves have no access.

It was more than a mere illustration when, in Boltzmann's day, increased entropy was equated with an increase in disorder, or even with the increase in unattainable knowledge or impossible foresight. Maximum entropy was compared with the loss of all structure, a state of chaos, or with the (relatively cold) "heat death" of the cosmos, i.e., with the uniform distribution of all molecules and the leveling of all temperature gradients. By the same token, a decrease in entropy was compared with an increase in order. And, in a certain sense, this already makes the birth of the stars possible, as a result of gravitation.

At the same time, the reversal of this decay, the transition to a homogenous mixture and to chaos - in other words the creation of order, of law and possible foresight - cannot be derived from Clausian thermodynamics. At best this ends at zero entropy, which can be exemplified by the condition of an ideal crystal at a temperature of absolute zero.
Entropy with a negative sign, on the other hand, can be conceived of as a measure of increasing order. This necessitates a full turn of the paradigm. For "measurable entropy exists in the case of energy because energy is homogenous" as Georgescu-Roegen remarks. At least, it can be thought of as being homogenous. He continues by saying that "solid matter", on the other hand, is heterogenous, as evidenced by the periodic table. Physicists have not made this transition. "Perhaps because western consciousness is governed by a mentality of affluence" as Georgescu-Roegen states (pp.304-305) and, of course, because the complex, the heterogenous has first to be analysed, "since thermodynamics does not pay heed to what happens with matter." At the end of the 1940s, Schrödinger was advised by B. Simon that he should not speak of order but of free energy, as it is the energy content in food, not the measure of order of matter that is decisive in maintaining the organization of animals (cited by Schrödinger on p. 80 of the 1969 edition). And this advice has in principle remained valid. (The above quotations are taken from the appendix of Jeremy Rifkin's book 1980; German edition 1982).

What, then, is negative entropy? This concept was obviously developed as a technical approach to the phenomenon of order. So let us inquire further:

WHAT IS ORDER?

I first critically examined this concept while working on *Die Ordnung des Lebendigen* published in 1975 (English: "Order in Living Organisms", 1978). I considered order in the sense of conformity to law, determination, constraints, or elimination of alternatives as well as in the sense of attainable foresight or obtainable instruction. Accordingly, I defined order as "law times the number of instances or applications."

This definition is based on the fact that even the most complex law is obviously incapable of establishing order unless it is applied as for example in the case of a state tax law that is so elaborate that it never leaves the drawer in which it was placed. On the other hand, a correlation as simple as that between mass and distance suffices to describe the gravitational conditions of the entire cosmos.

As far as the stucture of order is concerned, a hierarchical principle is found in almost all cases. In all matter, there are many hierarchical levels of structure, of substructures, superstructures etc. The most familiar instance of this are quantums, atoms, and molecules. The laws governing the individual hierarchical levels, however, differ from one another. In other words: the conditions governing both elimination and preservation - or the fundamental principles that enable us to correctly predict the order at hand - change from level to level.

The number of electrons in an atom depends on the quantum makeup or charge of the nucleus, while the type of electrons in an electron shell depends on their direction of spin, and the interconnection between neighboring atoms on the complement of the outer electron shells. We thus speak of nuclear charge, the Pauli exclusion principle, and chemical bonds.

We can get some idea of the increase in order by considering how much instruction or how many decisions would be required to assemble this type of matter from a construction set. It would be even better if we could get an idea of how many alternatives that are originally possible, but are excluded early in the process. It must be this very elimination of alternatives that enables information to be generated in this world.

Another example would be the increase in possible foresight or in confirmable prognosis that must be correlated with increased order when speeds in a plasma (as in the core of the sun) decrease to such an extent that the heavy particles - the protons - begin to capture electrons. At this moment matter is created. Knowledge of the location of a proton is commensurate with increased knowledge of the position of the electron, which had followed an unpredictable trajectory up until then.

If then hydrogen is burned to helium, then the predictability of the behavior and location of quanta will increase even further. This extends to the molecular level, from the simplest gas molecules to a carbon ring or even hemoglobin.

The instruction we describe in terms of a natural law represents a correlation in which the degree of increase of accurate prognosis must be interrelated with the alternatives eliminated by the natural law or with the constraints it specifies.

As early as the 1950s, Brillouin, Dancoff, Linschitz, Morowitz, and Quastler explored how many binary decisions would be required to select and correctly assemble all the essential atoms of an organism. For bacteria a value of 5×10^{10} to 10^{13} was determined, for humans 2×10^{28}. The latter figure presumably involves a greater number of binary decisions than contained in all the writings of man. And now to a second aspect:

THE QUALITIES OF ORDER

Structural hierarchy, a fundamental principle on the organism level, has not yet been taken into consideration in the above examples. It enters the picture as soon as we ask how many binary decisions would be necessary to piece together all the essential atoms of a human sperm. By any calculation the sperm must contain at least half of the instructions necessary to form a human being. The result: 10^{11}, the genetic material itself considerably less. Even if we accept 10^{11}, we must ask under what heading the missing 17 orders of instruction can be found (i.e., in relation to the above-mentioned 10^{28}).

I have formulated a clear answer to this question (Riedl, 1975). The 17 decimal places of order lie buried in the repeated applications of the same law in each emerging layer; in other words, they are hidden in redundancy.

We should keep in mind that we possess 10^{11} identical small gray cells in our brain, 10^{12} to 10^{13} identical neurons, 2.5 x 10^{13} erythrocytes in our blood, and that we produce 5 x 10^{15} such cells in the course of our lifetime; in the case of organelles and on the ultrastructural level the number of identical elements of all body cells reaches values between 10^{19} and 10^{21} (one sextillion).

This brings us back to our definition of order as law times the number of instances of applications. A law content on the scale of quadrillions times an application averaging in the sextillions leads to the undecillions defining the content of our order. In lower organisms the relation between law and application is shifted more in the direction of application. What we biologists describe as increasing differentiation in the course of evolution merely represents an increase in the law content and a relative decrease in redundancy. Even though the order remains intact in every case, this gradual reduction of redundancy - its transition into a specific conformity to laws - appears to us to be order of a higher quality.

The degree to which a law applies in the inorganic realm, of course, is reduced even further and the redundancy is greater (reaching an estimated level of 10^{80} identical quanta in the cosmos).

This transformation of redundancy to law content has an unexpected consequence, one that I would like to illustrate using a sequence of human artifacts. The laying of a mosaic can serve as our example.

Let us assume a total of 250,000 stones in 8 different colors (for a mosaic measuring 2.5 x 10 m). In order to sort them into sacks according to color, the helper needs to make only three binary decisions per stone. The rest is redundant: identical colors in the same sack. The apprentice, however, charged with laying a meander, must be given a great deal more instructions, even though the redundancy in the identical cycles of the meander will remain quite high. The craftsmen who put a quarter of a million stones together to depict the Battle of Issos (Museum of Naples) required an immense amount of instruction.

Nevertheless, a certain amount of redundancy remains even in this tumultuous battle scene - both for the craftsmen then and for the viewer today. Why? Because the artist, rather than stipulating the position of each stone, merely specified the coloration of structures and patterns. At the same time, the beholder need not discern each stone in order to recognize a horse, a rider, a helmet, or a tuft of feathers.

In our perception, the quality of order increases from the sacks of stones to the meander to the battle scene. In other words it increases with decreasing redundancy. Does this mean that order free of all redundancy is perceived as having the highest quality?

This, quite evidently, is not the case. In the case of architecture and music, Zemanek was able to demonstrate that our appreciation of harmony requires a certain mixture of both redundancy and surprise, of uniformity and variability. In music, for example, uniformity alone would be a stuck needle on a record; surprise alone would be mere noise. This is the first clue to the paradox ahead.

Redundancy-free order in the case of a mosaic would be a composition in which the position of no stone could be used to predict the location of any other stone. To complete such a mosaic, the craftsman would require an enormous amount of instruction encompassing the coordinates of each of the 250,000 individual stones. At the same time, the viewer would have the impression of total chaos. The mosaic would be indistinguishable from a product of pure chance and, a careful mixing of the stones would actually lead to the same result.

Transitions with such goals can be found in Surrealism, in the Viennese Fantastic Realism, in Escher and Hieronymus Bosch - wherever the cooccurence of characters which makes up objects we can recognize becomes unraveled. This breakdown is virtually complete in Art Informel, in the turmoil of tachism, and in similar tendencies in modern painting. And, as in tachism, where composition is left to chance, it is no longer possible to decide, in parallel cases, what is involved: a redundancy-free, qualitatively highest level of order or the lowest level of order (chaos).

This curious dilemma is an epistemological paradox. According to evolutionary epistemology, it has its origins in our cognitive apparatus (Riedl 1980, 1987) which is based on a spiral process of expectation and experience, of prognosis and confirmation. This, in turn, requires recurring events. Were God a one-shot influence - a non-recurring law through lack of repeated application - this alone would make it impossible for us to recognize him.

ORDER, INFORMATION, AND INSTRUCTION

Sometimes a certain degree of quantification has been put forward when referring to order, attainable knowledge, and predictability in the sense of Schrödinger's negentropy. This kindles the desire to apply metric methods more generally when describing levels and values of order. Shannon, in his celebrated study of 1949, was furthermore able to demonstrate that the average amount of information per signal in a language code could be expressed by the same algebraic formula that Boltzmann had developed in his formulation of entropy. Shannon even speaks specifically of "information entropy."

This prompted many other branches of science to employ the term entropy whenever sufficient predictability was lacking: the term low or negative entropy came to denote knowledge or possible information. This led Shannon (1956) to warn against this muddle. In an article titled "The Band-wagon", he cautions against blindly taking up ranks behind the marching band.

On the other hand, a certain similarity cannot be denied. Information scientists should by no means be obligated to master the Second Law of Thermodynamics, and thermodynamic experts should be allowed to stick to their expression free energy. At the same time, it is justified to take a closer look at this potential similarity. The question is an intriguing one and, as I hope to show, one that is probably decisive.

In a second volume, first published in 1976, I again focused on the problem of order. I sought something along the lines of an energy:information equivalent. I described the conditions of indetermination and determination in systems and the relation between attainable and unattainable knowledge. The subject matter involved the accretion of order in organisms and the discarding of entropy. When dealing with organisms, the order content is of greater interest than free energy; in the cases of their metabolism, which involves uptake and removal, the opposite appears to be the case.

While little can be gained by analyzing the free energy of an eye or of the optic lobe of the brain, one cannot overlook the degradation of order in metabolism - the life-maintaining process in the animal kingdom. I had thus again arrived at "feeding on order," a concept that, as remarked earlier, was not accepted by physicists even from the mouth of their Nobel-Prize winning colleague. And it became commonly accepted that free energy had no structural equivalent.

The 1980s afforded increasing opportunity to intensify research on evolutionary epistemology. In the wake of Konrad Lorenz's decisive treatise (1973), I developed the following perspectives in my contributions of 1980, 1985, and 1987 as well as in the book published together with Franz Wuketits.

Our innate mode of perception, it turns out, provides us not with a reflection of extra-subjective reality, but rather with the behavioral guides used by our ancestors to solve basic problems of life. And since the development and propagation of such behavioral adaptations throughout a species no doubt require over a million years, we are talking about quite early ancestors. These innate modes of perception are consequently adapted to solve the vital problems of predatory apes, at best of early man. These patterns are genetically fixed and cannot be changed - neither through training nor insight. Better insight can only help transcend them.

One has to concede that these simple, problem-solving guides are overtaxed by the increasing complexity of our civilization. This is at least clearly the case in the theoretical and abstract realm of free energy and negative entropy. We are dealing with sensory windows to the outside world. They frame key sections of extrasubjective reality and simplify their contents in qualitatively alternate interpretations. This is analogous to the familiar sensory qualities that simplistically present the world to us as a medium of color versus sound or of touch versus smell and taste. Nothing indicates the existence of a continuous wave spectrum. Instead, our sensory impressions are arranged to fit a color circle, and we have no clue as to the correspondence between the oscillations that reach our ears and our skin.

OUR COGNITIVE DUALISM

We have thus been provided with the expectation that this world embodies the alternatives of space and time, quantity and quality, as well as of causalities and finalities, that is to say, of alternatives such as cause and purpose or function. Within this context, attention should be drawn to the seeming alternative between static and dynamic states - an alternative we experience as form and function. How something is structured appears to us to be something completely different from its purpose. Consequently, we find the basic structure of verbs and substantives even in the most exotic languages and forms of communication developed by deaf-mutes.

It is more than peculiar that we distinguish between running and legs or between flying and wings in such an uncompromising and rigorous way, even though it is sufficiently clear that no animal ever flew without wings and no wing was ever designed without flight in mind. If form and function present themselves as disjunct rather than interrelated alternatives in the phenomenological world, then we must reckon with their preservation regardless whether we set our sight on greater or smaller categories. Our innate guide apparently compels us to extrapolate from this simplification, which was adapted to the perceptual problems on our moderate scale. Without hesitation we thus speak of force, which we experience in our biceps, even when reflecting on the powers holding the quanta of a nucleus or the stars of a galaxy together.

When contemplating our very existence or that of the cosmos, it should be of little surprise to discover form in the guise of body and matter, and function as soul and mind. We are distressed by this lack of unity because subjectively we regard ourselves to be a union of body and soul. The same holds true at the low end of the scale, where physicists joke that on Monday, Wednesday, and Friday quanta behave like corpuscles, and as waves during the remainder of the week. In the case of the quanta hoarded by Maxwell's demon, both descriptions or interpretations are applicable. The demon must have knowledge of the travelling quanta as objects, structures, or forms, however minute, i.e., information, or better instruction to the effect "the thing is here" or "the thing is not here". How one views the energy the demon uses to power his machine is another matter.

I suspect that a cognitive dualism lies at the root of these considerations and that a similar relation exists between particle and wave, form and function, body and soul, and matter and mind. This dualism is merely cognitive - a consequence of two manners of perception, or two sensory windows in a by no means divided world. At the very latest when confronted with the phenomenon of so-called quantum dualism, no doubt can remain that this dualism is a product not of suprasubjective reality, but merely of our senses.

Should my suspicion prove true, then the controversy surrounding energy and information can be viewed in a different light. It then follows that the interpretation of the 2nd Law of Thermodynamics in the sense of Clausius (as the loss of free energy) is one-sided. The world cannot be understood on the basis of forces, power, or transformation of energy or force alone.

There is simply no getting around Boltzmann's contribution, especially in Schrödinger's interpretation. The approaches attaching importance to information, instruction, structure, law, and order will no doubt turn out to be of greater significance. In my opinion the comprehension of this second perspective has become a question of survival for our energy-oriented science, industry, and society.

DO WE FEED ON ORDER?

Of course we do! Exactly as Schrödinger elaborated. To deny this merely because we cannot yet measure it would be disastrous. Indeed, the opposite is true. Our overindulgence is based on the very fact that we are at present unable to measure order.

In my opinion this is where the story becomes especially interesting. Why? Because we have left the realm of academic debate, the arena in which one school of thought is pitted against the other; the very survival of our species is at stake.

We humans exhibit a metabolism similar to that found in the animal kingdom. It is beyond doubt that we must continuously destroy order to merely maintain our existence. We consume legions of natural products - from apples to deer - in the course of our average 70-year life span. Approximately 90% of the energy is lost in the transfer from one level of the food chain to the next. Salad, a rabbit, or a pike - representing the first, second, and third steps in the chain - provide us with only a tenth, a hundredth, and a thousandth of the originally stored free energy.

We can get an idea of the enormity of this destructive process if we consider that each of those thousands of individuals involves the loss of 10^{20} to 10^{28} bits of instruction. Especially since our own metabolic wastes contain order of the lowest grade, and provide nourishment for unicellular organisms and bacteria only.

A discussion on this relationship between order and entropy in the biosphere started in the 1970s (e.g., Nicolas Georgescu-Roegen 1971, 1977 and Riedl 1973, 1973a, 1975) and, as mentioned earlier, Jeremy Rifkin devoted an entire volume to the entropy problem from this perspective in 1980 (including a review of the literature). Today, towards the end of the 1980s, we have been labeled "tendentious" and termed "green ideologists." Our colleagues have evidently still not grasped the severity of the situation. I would like to illustrate this taking examples from four levels of civilization.

Firstly: one generally overlooks the fate of the material carriers of free energy. Georgescu-Roegen states (in Rifkin 1980, p.304) that, "in thermodynamics, matter is referred to only in connection with the loss of available energy." "Yet the fact that available matter is likewise converted to unusable matter," he rightly admonishes, "is equally elementary and transparent." The recycling of the rubber worn off of automobile tires, is just as impossible as that of the phosphorous molecules in chemical fertilizer after it has been spread.

Yet we need not restrict our argumentation to the molecular level. The mountains of garbage produced by our cities belong to the same category of entropy, namely that level of chaotic mixture at which the expense of separation exceeds the expense of replacing the destroyed goods. Franz Kreuzer, Austria's Environment Minister in the mid-1980s, already recognized how this relates to Boltzmann's equation.

The story doesn't end here. Under the present conditions within industry, where the expenses (financial costs) of repairing an appliance is greater than producing a new one, the entropy principle is carried one step forward, i.e., the promotion of waste accumulation. A further level is attained when, on top of fully automated production, unskilled labor is incapable of anything else but delivering, replacing, and discarding (see Riedl 1988).

The conflict surrounding the preservation of a riverine forest in Austria (Hainburg) led to the establishment of the "Forum Österreichischer Wissenschaftler für Umweltschutz" (Forum of Austrian scientists for Environmental Protection). My appointment as president of the forum resulted in a professional cooperation with my friend Kreuzer: for example in the elaboration of an "environmental doctrine" for the Austrian constitution (preliminary publication, Aubauer 1987). A good deal of the following ideas stem from experience gained in the cooperation with that forum.

WE NOT ONLY FEED ON, WE PARASITIZE...

Secondly, we destroy more order than we create. This goes far beyond the unavoidable loss dictated by the 2nd Law, namely that the creation of order in an (open) system demands that an even greater amount of disorder is transferred to the environment. In the course of our production processes, we go as far as to destroy our capital.

It was in the study group "environmental doctrine" that the information scientist Gerhart Bruckmann coined the term "Brutto National Vermögen" (gross national assets). And, in fact, it soon became obvious that through growth of their gross national product, Austria and other industialized nations are in danger of losing several times more of such gross national assets.

For example, through the production of our steel and chemical industries we lose forests. It is apparent that in many cases the losses in assets exceed the gains through the products by orders of magnitude. The same no doubt holds true for the loss of soil quality due to chemical fertilizer and extensive agriculture: the assets here are quality of life and health.

The vagueness of my formulation is by no means based on any doubt about this correlation, but rather on the fact that we have, as yet, had no opportunity to determine our assets. What is a forest worth after three generations of management? Certainly more than the book value of the timber. The forest has conserved the soil, staved off erosion, landslides, and human resettlement, has purified the air, produced oxygen, and supported songbirds, game, and hiking trails. How can the value of this complex order be measured?

We are mistreating our forests precisely because we calculate in energy units - in monetary value and capital - rather than paying attention to their negentropy and order content. Instead of being helpful here, Clausius leads us astray. It is Boltzmann and Schrödinger who point in the right direction. We parasitize the assets that nature has provided us with. Consider the abandon with which we burn non-renewable natural resources such as the organic molecules of mineral oil.

Third: superimposed over the above are the destructive processes within culture - so-called cultural parasitism. Konrad Lorenz (1974) was the first to point out this phenomenon and we (Lorenz 1983 and Riedl 1988) have followed its development with interest.

In the organic realm there is but one strategy in which the reduction of order or de-differentiation has proven advantageous in the sense of reproductive success: parasitism. Neither dwarfism nor sense organ reduction in cave dwellers can serve as examples; in each of these cases this reduction is compensated by the refinement of other structures, especially sense organs.

The parasite alone is capable of increasing redundancy through de-differentiation - in the extreme case down to the attachment organs, fatty tissue and gonads. All this, of course, at the expense of higher order, i.e., decreased redundancy in the organization of the host. Only such a gradient of order can satisfy the requirements of the 2nd Law, and temporarily sustain life. For the parasite to profit in the long run, it must never endanger the survival of the host species - or evolution will eradicate the parasite with the host.

This decline of order finds its parallel in certain cultural processes of the modern age: the increased redundancy in modular and grid architecture and in urban planning, the trends from Art Informel to Dadaism and Tachism, a lyric that has progressively dispensed with rhymes, metrics, sentences and actual words.

The host is in this case our traditional cultural understanding - that value we naïvely associate with the harmony of a historic city, the elegance or cheerful appearance of a facade, a gallery, or a book whose title promises poetry. At the same time, this host - our values, our differentiated culture - should not be endangered. There is no guarantee that this "species" can hold up under the strain: at present we are hardly in a position to gauge what we are destroying. In the long run, the destruction of order can only be successful if it makes way to even higher order. It is too early to certify that this is the case. On the contrary, we have neglected this order, this negative entropy. Its destruction has therefore been all the more thoughtless.

...AND WE ARE DESTROYING THE EARTH

Fourth: we are destroying entire systems of order, even cultures. Entire cultures are being extinguished due to the same energy-minded utilitarian type of reasoning that has eradicated more modest strains of grain or potato.

Again, the economically and militarily, i.e., energetically, more powerful cultures prevail over the energetically weaker ones. No longer are the conquistadors, missionaries, development aid, ideologists, corporations and banks solely responsible for this destruction. Hypnotized by such power, the energetically weaker cultures contribute to their own demise (Riedl 1985).

Folklore and customs, tradition and language are sacrificed for Coca-cola, radios and washing machines. The energetic ideal of power in the form of power, watts, or capital - so easy to understand - has blinded them. Even we ourselves have difficulty resisting the pressure of expanding monocultures. In this case our defenselessness is promoted by ignorance, by our inability to gauge the value of law, order and differentiation. Especially when the energetic values are so simple and easily quantified. We should have taken Boltzmann and Schrödinger more seriously. We also have reason to fear that the cognitive dualism which prohibits us from understanding the systems interplay of static and dynamic states is rooted in our anatomy, namely in so-called laterality.

We know today that each hemisphere of our brain is specialized. While the left half is responsible for analytic, computer-like deductive tasks, the right attends to synthetic, creative, inductive tasks. The left is thus also responsible for the simplification to homogeneous processes and data, the right for the perception of heterogeneous, complex structuralization. At the same time, only the left hemisphere is directly coupled to speech and consciousness. Further development of Schrödinger's ideas may have been thwarted by this stumbling block.

In fact, even aptitude appears to be governed by hemisphere preference, whether by predisposition or environment. On might even suspect that those who predominantly use the left hemisphere may not be able to understand fully those who rely more on the right hemisphere, even to the extent of outright mistrust.

Considering that Schrödinger's negentropy concept is still met with "doubt and opposition" by physicists, I recommend as a biologist that all the above be born in mind. Not only will we fail to comprehend this world if we merely interpret it according to forces and not also in terms in order, differentiation, and instruction. We will eventually destroy it.

This as a footnote to a footnote of Schrödinger's, whose influence I have been able to elude as little as Schrödinger was able to elude Boltzmann's.

REFERENCES:

Aubauer, H. P. (ed.), *Umweltmanifest*. Unter Mitwirkung des Forums Österreichischer Wissenschaftler für Umweltschutz, Falter, Wien 1987.

Brillouin, L., *Science and Information Theory*, Academic Press, New York 1956.

Dancoff, S. and H. Quastler, "The Information Content and Error Rate of Living Things," in: Quastler, H. (ed.): *Information Theory in Biology*, Univ. Illinois Press, Urbana 1953, pp. 263 - 273.

Georgescu-Roegen, N., *The Entropy Law and the Economic Process*, Harvard University Press, Cambridge Mass 1971.

Georgescu-Roegen, N., *Energy and Economic Myths*, Pergamon Press, Elmsford-New York 1977.

Kerber, G., A. Dick and W. Kerber, *Dokumente, Materialien und Bilder zur 100.Wiederkehr des Geburtstages von Erwin Schrödinger*, Fassbaender, Wien 1987.

Linschitz, H., "The Information Content of a Bacterial Cell," in: Quastler, H., R. Augenstein, et al. (eds.), *Essays on the Use of Information Theory in Biology*, Univ. Illinois Press, Urbana 1953, pp. 251-262.

Lorenz, K., *Die Rückseite des Spiegels*. Versuch einer Naturgeschichte menschlichen Erkennens, Piper, München-Zurich 1973. English: Behind the Mirror. Methuen and Co., Harcourt, Brace and Jovanovich, London 1987.

Lorenz, K., *Die acht Todsünden der zivilisierten Menschheit*, Piper, München-Zurich 1973. English: Civilized man's eight deadly sins. Harcourt, Brache and Jovanovich, London 1974.

Lorenz, K., *Der Abbau des Menschlichen*, Piper, München-Zurich 1983. English: The waning of humanness. Little, Brown and Co., Boston 1987.

Morowitz, H., "Some Disorder - Order Considerations in Living Systems," *Bull. Math. Biophys.* 17, 1955, pp. 81 - 87.

Morowitz, H., *Entropy for Biologists*, Acad. Press, New York 1970.

Quastler, H.*(ed.)*, *Information Theory in Biology*, Univ. Illinois Press, Urbana 1953.

Quastler, H., *The emergence of biological organization*, Yale Univ. Press, New Haven-London 1964.

Riedl, R., Die Biosphäre und die heutige Erfolgsgesellschaft, *Universitas* 28 (6), 1973, pp. 587 - 593.

Riedl, R.: Energie, Information und Negentropie in der Biosphäre. *Naturwiss. Rundschau* 26 (10), 1973a, pp. 413 - 420.

Riedl, R.: *Die Ordnung des Lebendigen. Systembedingungen der Evolution*, Parey, Hamburg-Berlin 1975. English: Order in Living Organisms. A Systems Analysis of Evolution., John Wiley & Sons, Chichester 1978.

Riedl, R.: *Die Strategie der Genesis.* Naturgeschichte der realen Welt, Piper, München-Zürich 1976.

Riedl, R.: *Biologie der Erkenntnis.* Die stammesgeschichtlichen Grundlagen der Vernunft, Parey, Hamburg-Berlin 1980. English: The Biology of Knowledge. The evolutionary Basis of Reason, John Wiley & Sons, Chichester 1984.

Riedl, R.: *Die Spaltung des Weltbildes.* Biologische Grundlagen des Erklärens und Verstehens, Parey, Hamburg-Berlin 1985.

Riedl, R.: *Begriff und Welt.* Biologische Grundlagen des Erkennens und Begreifens, Parey, Hamburg-Berlin 1987.

Riedl, R. and F. M. Wuketits (Eds.): *Die Evolutionäre Erkenntnistheorie.* Bedingungen-Lösungen-Kontroversen, Parey, Hamburg-Berlin 1987.

Riedl, R.: *Der Wiederaufbau des Menschlichen.* Wir brauchen Verträge zwischen Natur und Gesellschaft, Piper, München-Zürich 1988.

Rifkin, J.,: *Entropy: A New World View.* The Viking Press 1980. German: *Entropie.* Ein neues Weltbild, Hoffmann und Campe, Hamburg 1982.

Schrödinger, E.: *What is Life? & Mind and Matter*, Cambridge Univ. Press. Cambridge 1943.

Shannon, C. and W. Weaver: *The mathematical theory of communication*, Univ. Illinois Press, Urbana 1949.

DETERMINATION AND SELF-ORGANIZATION: ERWIN SCHRÖDINGER'S VIEWS ON CHANCE

HERBERT HÖRZ
Institut für Philosophie
Otto-Nuschke-Straße 10/11, 1086 Berlin, Germany

1. FORMULATION OF THE PROBLEM

With his views on chance, Schrödinger became directly involved in the debate on determination and self-organization. Determination, in this context, refers to the conditionality and certainty of events. As N. Hartmann emphasized, it implies the "necessity of factors bringing about dependent events."[1]

Already, at an early date Schrödinger opposed the idea of absolute determination. Under the influence of F. Exner, Schrödinger developed his view of chance and of the statistical character of laws. Self-organization denotes the structuring of chaos, "the irreversible process which, through the interaction of subsystems, leads to more complex structures of the entire system."[2] Schrödinger is one of the thinkers who did the intellectual groundwork for a physics of self-organization in open systems.

The focus of this branch of science are the laws of non-linear cooperative processes far away from thermodynamic equilibrium. The fundamental ideas, in particular, originate from Schrödinger, Turing, Bertalanffy, Prigogine, Eigen, and Haken.[3]

Determination without chance is preformation, inevitable fate and implies the predictability of events. This is contrary to experience. Theories of self-organization require chance as a constitutive feature for structure formation. Above all, two of Schrödinger's contributions are relevant for understanding the relationships of determination and self-organization. By virtue of the fact that he founded and interpreted wave mechanics, he essentially influenced the discussion on the determinants of events. On deterministic positions, his views of chance contradicted the Copenhagen interpretation of quantum mechanics, because he insisted on the possibility of grasping in theoretical terms, events which could be visualized while emphasizing the continuity of physical processes. Together with Exner he distinguished the regular macrocosmos from the random microcosmos. But apart from that, he, the physicist, provided substantial incentives for the investigation of life processes with his answers to the question: *What is Life*? According to Schrödinger, life follows the principle of order from order. In the discussion on quantum mechanics, his views on the issue of chance had led him to regard the mechanical laws of physics as violated.[4] The stability of life processes is provided for by the statistical validity of the entropy theorem. Schrödinger remarks:

J. Götschl (ed.), Erwin Schrödinger's World View, 71–85.

What an organism feeds on is negative entropy. Or, to put it less paradoxically, the essential thing in metabolism is that the organism succeeds in freeing itself from all the entropy it cannot help producing while alive.[5]

Not only Schrödinger, but also his teacher Exner followed Boltzmann's way of thinking. To illustrate his views on physical laws as average laws, Exner cited the law of falling bodies and linked it with Boltzmann's considerations on fluctuations.

In conversations, Boltzmann completely agreed with this view and thought it not only possible, but very probable that the falling body moves along by fits and starts, perhaps not in a 'straight line' but in a zigzag line.[6]

According to Broda, Boltzmann succeeded in really understanding and interpreting entropy.[7] The idea of life as negative entropy was also attributed to Boltzmann. Referring to Schrödinger, Broda states:

Properly speaking, Schrödinger only formulated this and brought this to the attention of the scientific world. Later on, many physicists allowed themselves to be influenced by him."[8]

In my opinion, Schrödinger's theoretical and philosophical contribution lies in the fact that he emphasized the role of chance in a theoretical understanding of reality, at a time when physics was essentially still under the sway of mechanical determinism. With his views on chance he rejected the thesis of an unambiguous predetermination of events. In this respect, he did not adopt the position of indeterminists who only accept the random play of forces, while denying any regularities. According to Schrödinger, the macrocosmos is a system of elements that interact on a large scale on the basis of unambiguous laws; the behavior of elements (microcosmos) is random. Only this interpretation allows for the principle of macroscopic behavior and the individuality of living organisms, so that both the statistical interpretation of the second fundamental law of thermodynamics, and the stability of living organisms, can be explained. Thus, Schrödinger belongs to the theorists who understood the interrelation of determination and self-organization in terms of a philosophical notion of chance.

Theories of self-organization explain the transition from one structural and developmental level to another. Thus, the eternal existence of equal material structures as assumed in a one-sided view of determinism is rejected. The principle of development is opposed to the principle of the structural unity of the world. This latter idea was part of the philosophical interpretation of classical mechanics. The universality of mechanical laws demanded universal structural equality. In such a strict view of determination only such universal laws apply. In this context, the classical notion of causality as something necessarily bringing about a cause is identical with the regular course of events, with its predetermination and predictability.

The rejection of determination, an alternative to strict determination, often seems based on the assumption of pure chance. This assumption should explain why new things can emerge. In strict determinism, nothing new can emerge, since only what has been preformed can evolve as form. If we see development as the emergence of other, new and higher qualities through quantitative and qualitative changes of a basic quality (whereas the tendency for higher qualities to emerge is based on stagnation, regression as well as the deployment of all elements at a certain stage development), then self-organization is the basic form of development. Self-organization cannot exclude chance, because otherwise nothing new would emerge. But if we take into account

Schrödinger's argumentation, we see that self-organization is subordinated to regularities, because the development of a system is the interaction of various elements in the development process.

On the one hand, in theoretical terms, the rigid alternative to strict determination, i.e., lacking determination, disappears in dialectical determinism which grasps the conditionality and determination of objects and processes as the imposition of necessity on chance, as a random and necessary implementation of possibilities.[9]

Only when structure is understood in the sense of self-organization, is it possible to grasp the interrelation of being and emerging, and to direct controlled evolution towards specific relative objectives on the basis of evolutionary laws. An important issue in this connection is also the determination of self-organization, since the degrees of self-organization themselves prove to be degrees of determination.

If we see the self-motion of inexhaustible, objective reality as the unifying principle of the emergence of structures, the stability of structures and the dissolution of structures, then determination and self-organization are not opposites. However, in this context it must be emphasized that self-organization is only possible when determination is given together with chance. This is the scientific orientation of a search for internal potentials, causes and conditions for the emergence of new aspects. Self-organization proves to be an essential motive of determination, because the objective interconnection between objective aspects and processes is not universal chaos, but always in certain spheres under certain conditions represents a chaos which orders itself, and at the same time destroys order.

Thus, chaos turns out to be the first stage of order. On the one hand, this stage characterizes the dissolution of a certain structure, resulting in the end of the system with its specific system structures. But on the other hand, this first stage is the basis of structuring, hence the beginning of a new system in its evolution.

As was already emphasized, Schrödinger played an essential part in the theoretical elucidation of this interrelation. To make this even clearer, we should analyse more precisely Schrödinger's views on chance, taking into account his attitude to philosophy as heuristics, and locating possible principles of self-organization on the basis of his understanding of chance. This, in turn, will enable us to grasp contemporary views of determination and self-organization as a solution to philosophical issues which also play a role in Schrödinger's work.

2. LAW AND CHANCE

Laws of nature have an internal structure. They refer to possible events that are interconnected in a necessary and random manner. We are interested in this connection in conditioned chance which becomes reality with a certain probability. Law and chance form a unity, because even the dynamic law, i.e., the necessary realization of only one possibility, imposes itself as a tendency only in random events.

Schrödinger developed his views on chance from the positions of his teacher Exner, something he underlines several times. As he said in his inauguration speech at Zurich University December 9, 1922, it was he who, in 1919, for the first time criticized in philosophically clear terms, the matter-of-fact attitude with which everybody is convinced about the absolute determination of molecular events. He comes to the conclusion that what was asserted was indeed possible, but by no means necessary, and not very probable at all, if light is shed on it.

First, regarding non-necessity, I have already explained my views on this at an earlier date, and I think, with Exner that it can be maintained - in spite of the fact that most researchers are calling for quite definite features of elementary regularity. Of course, we can explain the energy theorem in great detail given the fact that it already applies on a small scale. But I do not see that we have to.[10]

Also in the work published in 1947, which deals with the particular character of the world view of natural science, Schrödinger remarked:

As Franz Exner already pointed out in his lectures published in 1919 (by Franz Deuticke, Vienna) which reduce the mechanical theory of heat to the statistics of numerous individual events, quite independently of the fact, whether the individual events constituting the material of statistics are in turn "causally" determined with absolute acuity, as had been assumed until then, or perhaps primarily in a random manner, with a considerable scattering of the success even under completely equal initial conditions. Exner assumed the latter case. It so occurred that the same assumption was put forth some years later by quantum mechanics and soon included in its credo. In this connection, Exner's name is never mentioned, probably because he had not published his ideas in a learned scientific journal, but in a readable textbook without formal trappings.[11]

Pointing to an important consideration of Exner, Schrödinger contended that the issue was not at all whether elementary events were causal or random, because the observed regularity was an implication of statistics with very great numbers. In fact this was really one of Exner's basic ideas.

What are the essential positions and arguments that Exner presented in his lectures? He saw mechanical laws as average laws whose validity is questionable in small areas. He cited examples such as the Brownian notion as evidence that laws have a limited validity downwards, because those elementary events are random on which the laws are based. In this direction, he also specified the law of causality. According to him, it expresses nothing else but the fact:

that the events of nature, as far as we can follow them in macrocosmic terms, i.e., in average values, occur not randomly but in a regular manner; whatever goes beyond this may be a necessity of thinking for many people, at least a convenience by simplifying the issue, but as was already noted, it does not mean that its accuracy automatically follows from the power of such a necessity of thinking.[12]

This notion of causality is a limited one. His only remark regarding this was:

there have to be some causes that not only determine the normal course of events but also guide it along regular lines.[13]

Exner then discusses the opposite of determination (coercion) and chance. Chance appears to be opposed to causality, and Exner put forth the task to establish an interrelation between causality and chance. In his opinion, chance is only possible in a multitude of options.

Each criterion for the random character of events is immediately connected with a multitude of them.[14]

In a single event, it does not make sense to speak of chance.

Probability considerations are not based on any assumption as to whether the individual events are causally conditioned or random. The only prerequisite is that individual events are independent of the others, so that the occurrence of one has no influence whatsoever on that of another.[15]

From the multiplicity and independence of the events, average regularity arises.

Exner was thus able to connect causality and chance by limiting their validity on two levels. The microcosmos supplies the elementary events which are random or about which no statements can be made. The macrocosmos consists of independent multiple elementary events with their elementary laws. Exner summarized his positions as follows:

1.We have distinguished between individual events and a multiple of them. 2. In the latter case, especially those multitudes interest us in which the individual events occur completely independent of each other; the question as to whether they are causally conditioned or random, we have excluded as irrelevant for further considerations. 3.We find ourselves in the microcosmos or in the macrocosmos, depending on whether we investigate the individual event in a multitude or the average of the latter. The random or regular course of events corresponds to these two possibilities.[16]

In this way, essential positions are justified which we find again in Schrödinger's studies. Exner had a uniform conception of the world in which the existing antinomies are taken into account,

people always come across antinomies, and this will not be any different in the future. One has to put up with these antinomies, but one must not want to explain them if one does not want to enter into sterile speculations.[17]

Exner saw seeming contradictions, such as those between causality and chance, reversibility and irreversibility etc., as antinomies. In this context, the contrary manifestations are attributed to different objects of cognition. Since all laws are statistical in character for him;

we cannot regard the previously quoted sharp opposition between the first and the second law as a dynamic or statistical law and thus also the contradiction between reversible and irreversible processes as being justified. We must only consider the mental objects that we call reversible processes as the limit unattained in the microcosmos, a limit which can be approached...through irreversible processes.[18]

In certain respects, Exner was fascinated by the distinction made between causality and chance - a distinction one finds time and again in the history of philosophy. Chance is seen as referring to what is not regular. Considerations on the role of chance in laws seemed less pressing. If this factor was, however, taken into account, it is then possible to apply the distinction between system behavior and elementary events which Exner made between macrocosmos and microcosmos, in order to more accurately grasp the internal structure of statistical laws. It then becomes obvious that a statistical concept of the law connects the necessary realization of a possibility of a system - under system conditions - with the random realization of possibilities of elementary behavior, the transition of an element from one state to another can be seen as probabilistic, because there is a transition probability for it.[19] This statistical notion of law is also compatible with philosophical interpretations of quantum mechanics. [20]

In this way laws encompass the necessary and random realization of possibilities under specific conditions.

Random events are possible events that can occur under specific conditions with a certain possibility. In other words: chance is an objective relationship between the infinite properties of an object, of a process or of a person (group), and between the infinite relationships of different events. This relationship is not based on the essential internal relationships of these interrelated components.[21]

We can distinguish between system-internal and system-external, between essential and inessential, between recognized and unrecognized random events. We can also distinguish between random manifestations of fundamental, structural, process-related and developmental laws.[22]

I would also like to point to two other interesting ideas developed by Exner. First, he indicated that the determinists who assume the absolute validity of laws always search for a "hitherto concealed mechanism" in order to explain "why and how these laws apply."[23] This approach was adopted for finding hidden parameters in a classical deterministic interpretation of quantum mechanics. Of course, the search for hitherto unrecognized regularities has not been abandoned altogether; only the orientation has been rejected as being compelled to always complement statistical laws in classical deterministic terms. In this context, Exner's warning applies: "We thus risk succumbing to a sort of physical mythology."[24] Second, Exner also continued to develop his unified conception of the world when he examined the relationship between nature and society with respect to regularities. Historical events, however, are not regular. Exner asks:

> Is Man not also part of Nature? And is not all human activity part of the events in Nature? Should not also the same tendencies prevail here? We almost could think, and indeed, wherever human activity occurs in terms of collectives, a tendency towards regularity manifests itself.[25]

Indeed, it is possible to justify the statistical conception of the law for society as well.[26]

In his basic considerations on the relationship between the law and chance, Schrödinger follows Exner's positions and arguments, and views chance as being intrinsic to a law.[27] In the relationship between system and element, statements on the determination of the elementary process are not necessary for the behavior of the system.[28] Statistical laws of Nature lose their rigid validity "to the extent that the number of cooperating individual processes decreases."[29] This means that the future cannot be completely revealed "through continually increasing precise observations."[30] Schrödinger made different statements on the issue of chance. He states for instance: "The ordered course of events that we encounter comes about is a manner which seems completely evident to us; one could say: by "understandable chance."[31]

An interesting form of chance stems from an intersection of causal chains. It can essentially influence our lives.[32] But people also speak of blind chance. Obviously, by "pure chance" he meant the independence of events as a basis of statistical laws, as formulated by Exner.[33]

Schrödinger was aware of the fact that quantum theory had given rise to a new situation in which the relationship of law and chance acquired new dimensions.[34] He responded with great enthusiasm to the attempt made by Bohr, Kramers and Slater to interpret in statistical terms the law of the conservation of energy. Bohr called it "a revolutionary experiment," Pauli spoke about the "reactionary Copenhagen putsch."[35] It failed to reach its goals and the antinomy of purely random element behavior and of the regular system behavior proved impossible to maintain. Later on, Schrödinger expressed the following view on quantum theory: "The laws of chance are not invalidated by this "invention", but their effects are modified"[36] - the modification itself being insufficiently reflected. This had to do with the fact that one clung to continuum physics, to the objective realities of waves, to the visual concepts formed by the classical mechanistic understanding.

The relationships between law and chance, as defined by Exner, are significant in a heuristic sense, in that they take into account the individuality of organisms which obey statistical laws and show relative stability. However, the considerations on individuality compel us to distinguish the regular course of events in life from the probability mechanism in physics. As he said: "We must be prepared to find a new type of physical law prevailing in it."[37] This is the principle of order from order which is not alien to physics as a principle, as Schrödinger emphasized. For the theory of self-organization, two essential ideas emerged. On the one hand, "a small, but highly organized group of atoms," we call the genetic code, is capable of producing orderliness in life processes.

Since we know the power this tiny central office has in the isolated cell, do they not resemble stations of local government dispersed through the body, communicating with each other with great ease, thanks to the code that is common to all of them.[38]

On the other hand, Schrödinger pointed out "that a prevailing order has the force to preserve itself and to cause ordered processes."[39] Since he considered living beings to be open systems, he was able to find the metabolism "by which an organism maintains itself in a stationary manner on a rather high degree of order (a rather deep degree of entropy). "This metabolism," in reality, consists of a continually "sucking in" of orderliness from its environment."[40]

3. PHILOSOPHY AS HEURISTICS

Heuristics implies thinking beyond the confines of a specific discipline, a reflection on the philosophical, epistemological and methodological problems of one's own research work, and an examination of axiomatic formulations as well as necessities of thinking in their arguments. Schrödinger followed this approach as can be seen in his considerations of law and chance. He was criticized by Bridgeman for emphasizing the interrelation between science and human culture and calling scientific discoveries meaningless beyond their cultural framework. However, Prigogine/Stengers quote him in a positive sense, so as to emphasize the interrelation between concept-oriented development in science, and the cultural problems of a specific time period.[41] Certainly in this respect, there is no precise determination of science by culture, since natural science itself has a cultural function. It is, however, interesting to investigate more closely the interrelation between mankind's stages of culture in its development and concrete historical types of science.[42]

Schrödinger pointed out how important it is when special problems arise in a given theory, to grasp the arising antinomy in its historical roots in order to find the right solution to a given problem. In his study *Die Natur und die Griechen* (Nature and the Greeks), he justified historical retrospection with good reason. He characterized the split in the physicists' camp in terms of world outlook.

If one dares divagate to the metaphysical aspect of the object in a physical or biological publication, even in a popular one, then this is considered as improper. The natural scientist who does so must expect to be rapped on the knuckles; often it remains unclear, be it because of an offence against natural science or against the special metaphysical doctrine advocated by the critic.[43]

Certainly philosophy is hardly able to solve a specialized problem of a specialized science. If a philosophically ascertained research program reaches its goals, then philosophy appears superfluous. However, whenever universal enigmas are to be solved as part of a research program, experiments are waiting for theoretical clarification, crisis situations emerge in scientific disciplines, philosophical analysis is in demand. Then philosophy has to provide proofs to explain universally recognized scientific insights, by fitting them into a picture of the world. However, it also has to help justify new ideas and to provoke new questions, thus initiating a growing awareness of the problem. Philosophical knowledge, be it of a historical or systematic nature, is part of the cultural background for creative achievements in science. Therefore, it is necessary to avoid the "grotesque phenomenon" described by Schrödinger in the following terms:

...scholars of natural science, who are highly educated have an incredibly undeveloped or atrophied philosophical outlook of the world.[44]

This represents a challenge for education. It is interesting to note, if one pursues Schrödinger's considerations on Greek philosophy, that he prefers the views of Democritus to those of Epicurus. According to him, Democritus is aware of the fact that he does not know everything, whereas Epicurus is convinced of knowing fairly everything. Besides, Epicurus also added some nonsense to the system of Democritus.[45] Certainly Schrödinger's criticism is justified. It is only striking that he does not comment on Epicurus' views on chance. For a long time, they were also considered a faux pas in philosophical historiography, a gaffe which eroded the determinism of Democritus, because Epicurus admitted deviations of the atoms from the orbit which was predetermined by fate. But since he introduced chance as a constituent part, on the basis of the assumed declination of atoms, into the explanation of nature, he revealed that changes and developments are only possible through interactions, that new things can only emerge if the laws of fate are broken.

Schrödinger also pointed to the problem of free will:

By the laws of physics we are forced at every moment to do whatever we do. What is the point then in considering, whether it is right or wrong? Where is there room for a moral law, if the omnipotent law of Nature does not give that one a chance to speak? Today the antinomy is as unresolved as it was 23 centuries ago.[46]

In Epicurus' writings, the recognition of chance is directed against the view of predestination by fate, against fatalism. Because if everything were necessary in the sense of an unambiguous predestination, then freedom would in fact be only ignorance, he argued. However, if, given the objective existence of chance, freedom is increased through greater knowledge, man can put his knowledge of the objective laws to ever better use, so as to attain the desired goals. In this case what Epicurus says applies: "For the wise man, chance has ever less effect."[47] Schrödinger solves the dilemma with his philosophical view of the relationship between law and chance. According to him;

the spatio-temporal processes in the body of a living being which are in line with its intellectual activity and with its consciousness, or actions carried out in whatever way, are deterministic in a statistical sense, if not strictly deterministic in nature...[48]

I would prefer to speak of dialectical determinism which distinguishes causality and law from one another. Here, causality as such is one manifestation of the interrelation of objects and processes implying an interaction of systems (effects) by internal or external factors (causes). Chance or necessity of a causal relationship always appears in a complex network of causal relationships. In this context, existing groups of conditions are the basis for a range of possibilities with regard to further events. If a possibility is realized, then we can only speak of the "initial cause" of an "effect", if the causes for the existence of the possibility (coming true later) can by necessity lead to its becoming materialized. Other realized possibilities also have certain causes which did not become the initial cause of the final effect.

As opposed to causality, scientific laws cover universally necessary, that is, reproducible and essential interrelations between objects, which determine the character of a manifestation. Hence, it is not sufficient, as Schrödinger shows, to associate chance occurring on a massive scale in microscopic terms with macroscopic regularity. System laws provide a framework for random behavior allowing the range of random variations to determined. This can be seen in the probability of conditionally random causes when possibilities of elements within a system are realized.

If one advocates such a dialectically deterministic concept - and here Schrödinger's statistical deterministic views fit in, - then it is also possible to elucidate the problem of freedom. Freedom is the realization of a certain possibility determined as the goal of action, and this realization is based on competent decisions. The discrepancy between the goal and the results is due to both insufficient special knowledge and lacking social force, both decisive impeding conditions and opposing vested interests. The potential increase in freedom is decisive. This freedom can be obtained by greater domination of the natural and social environment and behavior to ensure well-being and to achieve meaningful goals.

Schrödinger tried to solve this problem by interconnecting the following premises:

1) My body functions as a pure mechanism according to the Laws of Nature.
2) Yet I know, by incontrovertible direct experience, that I know that I am controlling its movements of which I foresee the effects that may be fateful and all important: in which case I feel and take full responsibility for them.[49]

From these two premises, he draws the following conclusion:

I - I in the widest meaning of the word, i.e., that is to say, every conscious mind that has ever said and felt 'I' - am the person, if any, who control the "motion of the atoms" according to the Laws of Nature.[50]

This confirms our general claim again: Determination does not exclude self-organization. Self-organization also encompasses the control and regulation of events by human beings, which leads to the social evaluation of actions, taking place on the basis of values which are meaning-oriented relationships of facts for Man, e.g., usefulness, morality and beauty.

Reflections on autopoiesis generally move in this direction. H. R. Maturana justified determinism as an epistemological necessity, but did not conclude from this that we cannot always predict the changes in state of the investigated systems:

But a prediction is calculation within a system in a range (a model), and it is said that this prediction is isomorphic with a system in another range where the observation of the predicted events will take place. Therefore the fulfillment, or non-fulfillment, of a prediction is always a function of the relationships which the observer establishes between both ranges or spheres and is not independent of them in one of the participating systems.[51]

The principle of interaction, of purposeful change, also plays a role in Schrödinger's thought, when he speaks about determinism and free will. According to Maturana, freedom motivates the extension of our interactions beyond the limit of our spheres of structural coupling, whereby man as an observer can avoid the reduction of creativity and of novelty.[52] Indeed, it sometimes seems as if autopoiesis as self-production leads to self-contained interacting systems, and therefore autopoeisis is considered an explanation of the autonomy of life.[53] But if the independence of the systems is to be emphasized, because the interaction of systems is not simply interaction between all elements of a system, as could also be suggested by the thesis of the existence of open systems, then in my opinion it is legitimate to assume the independence and autonomy of the systems. Insight into the openness of the systems (cf. Prigogine), into the relative closedness of the systems continues to be an essential foundation for theories of self-organization. Otherwise, a modern monadology would have to clarify the problem of preestablished harmony.

In accordance with the heuristic thrust of general philosophical considerations on the relationship between law and chance, Schrödinger, the physicist, sought to explain biological phenomena and to answer the question as to what life is from the view point of a physicist thinking in statistical terms. In his philosophical view of objective reality Schrödinger agreed with Einstein. Both advocated field physics, continuum physics, although Einstein belonged to the founders of quantum theory. Above all, both of them considered probability statements to be problematic. Although it is possible to understand probability as a measure for the realization of certain possibilities and thus also as a measure of what is random, Schrödinger emphasized:

> It seems to me that the concept of "probability" is often abused nowadays. Probability also includes a statement as to whether something is or is not, even if expressing a certain doubt. Yet this is also only meaningful if one is convinced that the something involved is quite certainly or is not. A probability statement presupposes the complete reality of its object.[54]

Einstein endorsed this viewpoint in a letter to Schrödinger:

> (Apart from Laue) you are the only one among the contemporary physicists to see that one cannot avoid the positing of reality - if one is only honest.[55]

Einstein liked Schrödinger's thought experiment involving radioactive disintegration in which a cat locked in a box is neither alive nor pulverized, in keeping with the Copenhagen interpretation of quantum mechanics being considered at that time. This "cat half dead, half alive" was to indicate that quantum mechanics and its probability interpretation do not provide a complete description of reality. Einstein comments on this:

> Is the state the cat is in to be produced only by the physicist who is investigating the thing at a determined time? But in truth, nobody has any doubt that the existence or non-existence of the cat is something independent of the act of observation.[56]

The epistemological problem consists in the fact that reality is always a possibility that has been realized, but at the same time it tends to change further and to develop as new possibilities are inherent. In normal understanding there are only two possibile states for the cat: life or death. The probability is reduced to zero or one. But if we investigate ongoing processes with regard to their result, more possibilities exist, each with a certain probability. A field of probabilities emerges which is reduced when one possibility is realized.

This idea becomes significant above all, if the possibilities characterize the possible intermediate forms of alternative states. The more complex the systems, the more important the differentiation of the field of possibilities. This applies to possibilities of social changes, to the behavior of social groups and of individuals. In the course of events, understood as a process, possibilities that are realized in a conditionally random manner have a certain probability as a degree of coming true. But, by the result of events, only one possibility has been put into practice. Now, if we proceed from the assumption that this possibility necessarily realized owing to the entirety of conditions, then we must also consider the infinity of the objects and processes that do play a role in this respect. Therefore, if from the necessity of the course of events, we derive the necessity of all parts of this course of events, then we contradict the very view of chance introduced by Schrödinger himself. Now as long as Schrödinger insists on his view of reality in which states always exist unambiguously, then the conditionally random materialization of possibilities with a certain probability in the process of events is denied, and only the result of the course of events is taken into account. Insisting on the existence of objective chance, he was also forced to recognize probabilities of transition and of realization. This antinomy in Schrödinger's thought seems not to have remained unresolved.

4. PRINCIPLES OF SELF-ORGANIZATION

Contemporary thinkers have reflected on the philosophical implications of the theories of self-organization.[57] This is absolutely necessary, as further developments and theories on self-organization entail new epistemological and methodological questions. Now let us proceed from the previous considerations on the relationship between determination and self-organization in connection with Schrödinger's views on chance. In this context, an attempt is to be made to turn attention to essential philosophical principles which are part of the foundations of a theory of self-organization.

First: *There are no eternal structures of matter, only a continuous transformation of properties.* Time proves to be such a form, i.e., the existence of matter which expresses the duration, direction and order of the course of events. Duration encompasses the structural stabilities of a system from the emergence of structure to the dissolution of structure. Order indicates the aspects of structural changes appearing at some point or other. In this context, there exist intrinsic times of a system which can be projected to normal time. The cyclicity of the proper times is ordered, if state C follows state B which is conditioned by state A. The order of earlier and later aspects in the cyclical course of events, however, also allows transition from state C to state A and thus to the starting point of the development cycle. By means of the developmental criteria, it is possible to prove that the return to the old state is only seeming, because a new or higher property was reached. Only then can time be seen as being directed in one sense.[58] Self-organization expresses the emergence in the course of events, the appearance of fields of possibilities and their realization, that is, also the field of possibilities for the emergence of fields of possibilities.

Second: *Self-organization is based on the autonomy of systems which is founded on the cooperation between the elements of the system.* Systems with their own system laws interact among each other. Each system has its own system laws which have a statistical character, because they cover the interrelation between a necessary tendency of behavior and a conditionally random realization of possibilities of elements with a probability. Because of its internal possibilities, every system has a field of possibilities for further behavior. By the interaction of systems, changes in states result which are contained as a potential in the autonomy of the systems. Thus, interaction is the realization of possible states dependent on internal and external influences. Hence, on the one hand, the autonomy of the system must be taken into account, because it is only through their autopoiesis that the field of possibilities is determined.

External influences are modified and processed by the internal conditions of the system, provided that the system is not destroyed. On the other hand, the autonomy of the system does not transform the system into a monad, nor into a self-contained entity. Interaction becomes an essential aspect of the very existence of the system.

Third: *In self-organization there always exists a structural hierarchy which is expressed both in the master-slave relationships and in the existence of various structural and developmental levels.* We can consider physics as a theory of possible inorganic behavior. Then the field of possibilities thus constructed also includes possibilities for biotic evolution. Schrödinger has always pointed out that biological statements must not contradict physics which includes possible developments. The distinctive features of life do not consist in new physical principles, but in restrictions of the field of physical possibilities. In this context new possibilities of behavior appear. Restrictions stem from the different type of cooperation between the elements and from qualitatively new forms of self-organization. This also applies to chemical behavior. Possible physical reversion processes are directed only temporarily and unilaterally. If in this context we point to the interrelation between lower and higher developed forms of motion in which the theory of the lower forms of motion becomes the theoretical framework for the behavior of the higher form of motion, then an interesting connection between scientific theories results. Biological theories also prove to be a theoretical framework for social behavior. This is in fact the case as far as we are concerned with the genetic-biotic predispositions of individual behavior. However, this also applies to the genesis of social and moral behavior. For social biology an interesting field of activity opens up here.

Fourth: *Self-organization in developmental processes leads to relative goals in the course of events.* Development as a tendency towards the emergence of higher properties which imposes itself through stagnation and regression, and through deployment of all elements of a developmental stage, encompasses in its cyclicity apparent and genuine progressions as well as apparent and genuine regressions. The apparent return to the previous stage, to its existential and organizational forms, do not necessarily mean a genuine regression, if the functions and the initial quality of the developmental cycle are fulfilled in a better way, in qualitative and - to a greater extent - in quantitative terms. By contrast, an apparent progression in social change can very well mean a genuine regression and this fact may sometimes be concealed by reforms. Relative goals in the course of events sometimes result from programs, such as the genetic program, from the fields of possibilities of the laws and from the tendencies of further development. Their relativity sometimes consists in the very fact that a field of possibilities exists, and not only one possibility alone. To put it in terms of Schrödinger's mental experiment: the cat cannot only live or be alive, it can live euphorically or apathically, it can pine away or receive new courage to live, it can be dying but not dead yet. It is possible to justify relative goals in the objective course of events by irreversibility as a fundamental property of matter, by the existence of fields of possibilities and by the evaluated self-organization of systems. They are manifested in a state of order by fluctuations.[59]

Fifth: *Self-organization also implies that the future is open.* Contrary to the views of classical determinism, the future is not preformed by the present. Objective laws in their fields of possibilities also only determine the framework of future behavior, the range of variance of the future course of events. This does not exclude the existence of essential tendencies which impose themselves, given the appropriate conditions. It is this fact that enables creative behavior and freedom. Scientific predictions are possible, but do not affect the future course of events in an unambiguous manner. However, it can be assumed that they indicate possibilities which are realized in a random manner with a certain probability. On the basis of competent decisions, man can intervene in the conditional structure for laws, and thus stimulate or impede the conditions for putting some possibilities into practice. The risk he takes in this context consists of a regular part which accounts for the difference betweeen the certitude and probability of a possibility being realized, and the risk of behavior which involves his attitude towards a regular risk, his courage to take risks or his hesitation. Not always is it possible to know what could happen; it is thus always important to ask oneself the following question: what is to be decided when? The self-organization of society is always linked with decisions, conflicts of interests and contradictory acts performed by humans. However, the parallelogram of forces indicates a resultant in terms of which the result of self-organization can be seen as real history.

Sixth: *Self-organization in society calls for human control and regulation of the course of events.* Man is responsible for controlling his own activities so that their implications are not detrimental to mankind and serve the benefit of man.[60] The greatest consequence for man is a gain in personal freedom on the basis of social progress under peaceful circumstances. Social progress means improving the quality of life, not simply growth. Technological development, ecological problems, the danger of a global war with weapons of mass destruction, compel us to proceed from the responsibility for the implications.

The principles mentioned here constitute a possible framework for theories of self-organization. They direct our attention to the totality of events, the unity of nature and society, science and culture, man and technology. Self-organization is part and parcel of the self-motion of matter and the eternal change of form. Theories of self-organization can lead to specific philosophical statements, which give developmental theory a materialistic dialectical foundation.

5. SUMMARY

With his views on chance, Schrödinger contributed to the philosophical controversy on the relationship between determination and self-organization. His principle of order from order gives heuristic leads for further research. Indeed, the statistical interpretation of the law of the preservation of energy proved to be problematic, whereas the statistical interpretation of the law of entropy showed that the principle of the stability of living beings is a reception of negative entropy. Schrödinger's philosophical position contains antinomies, such as the recognition of objective chance, and this position is connected with his views on reality, views in which emergence, i.e., the conditionally random realization of possibilities with a probability, is denied. But Schrödinger's investigations are an important step to be followed by more comprehensive theories of self-organization which do not exclude, but rather include determination. Thus even today, theoretical development in the field of self-organization which has been further differentiated turns out to be a major challenge to philosophy. As philosophers and scientists, we should follow the example of Schrödinger the physicist who did not hesitate to address burning philosophical issues

NOTES

1 N. Hartmann, *Der Aufbau der realen Welt*, Berlin 1940, p. 310.
2 W. Ebeling; R. Feistel, *Physik der Selbstorganisation und Evolution*, Berlin 1982, p. 11.
3 ibid.

4 A. Hermann; K. v. Meyenn; V. Weisskopf (eds.), *Wolfgang Pauli. Wissenschaftlicher Briefwechsel mit Bohr, Einstein, Heisenberg, et. al.*, vol. I: 1919 - 1929. New York, Heidelberg, Berlin 1979, p. 70.

5 E. Schrödinger, *What is Life?*, Cambridge University Press, New York 1967, p. 76.

6 F. Exner, *Vorlesungen über die physikalischen Grundlagen der Naturwissenschaften*, Leipzig - Vienna 1922, p. 670.

7 *Ich bin also denke ich. Die evolutionäre Erkenntnistheorie. Franz Kreuzer im Gespräch mit Engelbert Broda and Riedl*, Vienna 1981, p. 27.

8 ibid.

9 H. Hörz, *Der dialektische Determinismus in Natur und Gesellschaft*, Berlin 1975.

10 E. Schrödinger, *Was ist ein Naturgesetz? Beiträge zum naturwissenschaftlichen Weltbild*, Munich - Vienna 1967, p. 14.

11 ibid., p. 49.

12 F. Exner, op. cit., p. 674.

13 ibid., p. 676.

14 ibid., p. 688.

15 ibid., p. 689..

16 ibid., p. 693.

17 ibid., p. 725.

18 ibid., p. 712.

19 H. Hörz, *Zufall. Eine philosophische Untersuchung*, Berlin 1980.

20 H. Hörz, *Atome, Kausalität, Quantumsprünge*, Berlin 1964.
H. Hörz, *Werner Heisenberg und die Philosophie*, Berlin 1966.
U. Röseberg, *Determinismus und Physik*, Berlin 1975.
U. Röseberg, *Quantummechanik und Philosophie*, Berlin 1978.

21 H. Hörz, *Zufall*, loc. cit., p. 84.

22 H. Hörz, *Der Dialektische Determinismus in Natur und Gesellschaft*, loc. cit., pp. 196 ff.

23 F. Exner, loc. cit., p. 714.

24 ibid., p. 709.

25 ibid., p. 707.

26 H. Hörz, *Zufall*, loc. cit.., p. 191.

27 E. Schrödinger, *Was ist ein Naturgesetz?*, loc. cit., p. 10.

28 ibid., p. 11.

29 ibid., p. 12.

30 ibid., p. 26.

31 ibid., p. 49.

32 ibid., p. 53 f.

33 ibid., p. 55.

34 E. Schrödinger, *Was ist Leben?*, loc. cit., p. 26 f., p. 41.

35 U. Röseberg, *Szenarium einer Revolution. Nichtrelativistische Quantenmechanik und philosophische Widerspruchsproblematik*, Berlin 1984, p. 119 ff.

36 E. Schrödinger, *What is Life?*, loc. cit., p. 97.

37 ibid., p. 86.

38 ibid., p. 85.

39 ibid., p. 84.

40 ibid., p. 79.

41 I. Prigonine; I. Stengers, *Dialog mit der Natur*, Munich 1981, p. 25 f.

42 H. Hörz; U. Röseberg, "Wissenschaftsentwicklung als Moment welthistorischer Kulturentwicklung," in: DZfPh issue 8, 1987.

43 E. Schrödinger, *Die Natur und die Griechen*, Hamburg 1956, p. 18.

44 ibid., p. 19.

45 ibid., p. 103.

46 ibid., p. 122 f.

47 H. Hörz, *Zufall*, loc. cit., p. 30 f.
48 E. Schrödinger, *Was ist Leben?*, loc. cit., p. 30 f.
49 ibid., p. 123.
50 ibid., p. 123.
51 S. J. Schmidt (ed.), *Der Diskurs des radikalen Konstruktivismus*, Frankfurt a.M. 1987, p. 115.
52 ibid., p. 116.
53 ibid., p. 119.
54 K. Przibram (ed.), *Briefe zur Mellenmechanik*, Vienna 1963, p. 34.
55 ibid., p. 36.
56 ibid., p. 36.
57 Cf. the extensive list of literature in: D. Ebeling; R. Feistel, *Physik der Selbstorganisation und Evolution*, loc. cit., pp. 429 ff.
58 H. Hörz, *Philosophie der Zeit*, manuscript.
59 H. Hörz; K. - F. Wessel, *Philosophische Entwicklungstheorie*, Berlin 1983, p. 118.
60 H. Hörz; D. Seidel, *Humanität und Effektivität - zwei Seiten der wissenschaftlichen-technischen Revolution?*, Berlin 1984.

SCHRÖDINGER, THE SELF AND THE GENES

WERNER LEINFELLNER
Department of Philosophy, University of Nebraska
Lincoln, NE 68588, USA

1. SCHRÖDINGER'S IMPACT ON BIOLOGY AND PHILOSOPHY

In 1948 E. Schrödinger, the physicist and internationally renowned Nobel prize winner, surprised the scholarly world with a small book on biology and genetics titled *What is Life?* In this book Schrödinger deals with the most important scientific breakthrough in modern times and maybe in the history of man: the discovery of the DNA or the genes. He analyses the function and significance of the genome (the genes) which determines the nature of every living organism from the amoeba to man. In his fascinating last chapter he points to the philosophical consequences of this new genetic knowledge.

Physicists had already begun taking an interest in biology in 1930, when Niels Bohr speculated on the biological implications of the uncertainty principle and the new concept of complementarity. Delbrück's and Avery's ideas about the physical nature of the genes inspired Schrödinger to write his small book which became a major influence on biological research and philosophy of biology. It motivated a number of physicists and philosophers to leave their traditional fields and to turn to the molecular biology and genetics. What made Schrödinger's book so fascinating was the lucid way in which he dealt with the gene. The gene, for him, was not an abstract scientific concept, but a concrete chemical and physical unit, the carrier of life. His ideas stimulated scientists and philosophers to think about biology and genetics in a completely new way. Schrödinger was the first to suggest that the gene could be viewed as a carrier of information with a physical structure which contained a succession of information in a hereditary code script. With his enthusiasm Schrödinger was able to convince physicists, biochemists and philosophers. I had the opportunity to meet Schrödinger several times after his return to Vienna in the years 1956-60, and I will never forget his prophetic words: "The first half of our century belonged to quantum physics, but the second half will belong to molecular biology and genetics. We have reached a point of dramatic change in our views of life and ourselves, the most important breakthrough in biological research with great discoveries is imminent. Its implications will change our culture."

Indeed, his book has influenced our modern conception of genes, of genes as order-sustaining factors for living organisms, but nolens volens, his new ideas opened a pandora box of paradoxes and riddles, for example, about entropy and the philosophical consequences of the new discipline which he called molecular biology. A revolution in biology and genetics began which has not yet ended.

Strangely enough, Schrödinger, one of the most famous physicists of all times, tells the reader on the first pages of his book that physics and chemistry are unable to explain what happens within a living organism. According to him, new laws are at work in the organisms and genes.

J. Götschl (ed.), Erwin Schrödinger's World View, 87–98.

Thanks to biological and genetic research we now know that living organisms are governed by higher laws that cannot be reduced to the ordinary laws of physics. It was Schrödinger's greatest merit to base life on these higher and complexer laws, higher than those of the simple, natural sciences.[1] These higher laws, neither neovitalistic nor immaterial in essence, are different from everything scientists have ever tested and described in their laboratories. They are metalaws similar to the symmetry laws, the TCP-theorem, the conservation laws, and exist only in living organisms. In these organisms they are superimposed on the natural laws. As scientific laws, they do not contradict but modify the laws of classical and quantum physics, acting upon physical and chemical matter through the genes or hereditary substance. In 1982, Prigogine pursued Schrödinger's idea of superimposition in proposing, for example, that in living organisms a special bio-operator "A" is superimposed on the quantum physical operator ψ[2]. Like the laws of symmetry, Schrödinger's higher laws are complex, evolutive and dynamic, and explain the increasing order, stability and complexity of living organisms.

Schrödinger's concept of the relative stability of living organisms anticipated the most important notion in today's biology and evolution theory, i.e., evolutionary stability, recently defined by Maynard Smith and Leinfellner.[3] Schrödinger was the first to assume that the increasing stability of organisms did not suddenly appear out of the blue, but had already evolved in the inorganic lifeless world. His famous example is the increase of order and stability in crystals. Given primitive forms of order in prebiontic systems, organic systems can evolve out of these already existing forms of order. This means, in Schrödinger's own words, "order is based on order"[4], or that higher order is always based on lower order. Modern biologists and physicists have taken up Schrödinger's idea, proving that the evolution of life is an optimization process of order and stability, in spite of the second thermodynamic law of entropy. Today, Prigogine and Onsager have removed the last obstacles in Schrödinger's gigantomachy vis-a-vis entropy or the decay of life through entropy and thermodynamic equilibrium. This will be shown in greater detail in chapter 4.

Schrödinger's novel ideas can be summarized in the following way:

1) There is a tendency in our cosmos or, at least in a part of our cosmos, for systems to move towards order and stability and to show an increase in their order and stability. This tendency, already existing in prebiontic form in inanimate nature, has become a precondition of the evolution of living organisms.
2) Laws sustaining a higher order function only in living organisms through the genome. The genome is responsible for the relative stability and ensures the maintenance of order in all organisms throughout their lifetime, inheriting this property to the progeny.
3) With the help of the genome higher laws moderate, weaken and even suppress the irreversible deadly tendency of all lifeless physical systems, including the universe, to increase disorder, destruction and inertness.
4) Schrödinger foresaw very clearly that the higher laws only function when the law of entropy is not violated, but was not quite clear how this works in detail. (See chapter 4.)
5) His conception of the stability of the human genes (genome) supported his idea that the ego, or what we traditionally call our soul, has been brought into existence and functions through our genes. In brief, genes determine the stability of our personality. Since the genes remain stable throughout our lifetime they ensure that our self-identity is maintained. One san say he regarded our ego, the soul which is traditionally regarded as being eternal, as finite and empirical, situated below Freud's and Jung's psychic level of the unconscious. This led him

to formulate original philosophical implications about consciousness, our soul, the self and the 'I', which are - as he himself underlined - alien to our Western thinking, but closer to Schopenhauer and Indian Vedanta philosophy. (See chapter 5.) Strangely enough, it was the rational physicist Schrödinger who initiated the exodus of "discontents" (with our culture, according to Freud) from our traditional western ideology. His book triggered the lingering suspicion that the world view of modern biology and physics is inconsistent with our traditional Christian culture (but not our Christian ethics).

2. SCHRÖDINGER'S FIRST OBSTACLE: THE LAW OF ENTROPY OR HOW LIVING MATTER EVADES THE DECAY TO EQUILIBRIUM AND TO ENTROPY

For Schrödinger the most striking characteristic of life, is the organism's tendency to escape the deadly state of thermodynamic equilibrium or of maximum entropy. He asks again and again how this is possible. Schrödinger regarded himself as "statistical determinist" like Boltzmann who had shown that the second law of thermodynamics is a statistical law. Its affirmation that certain processes do not occur does not mean that they are impossible, but merely that they are extremely unlikely. In microscopic systems the second law is violated regularly, but in macroscopic systems, which consist of vast numbers of molecules, the probability that the total entropy of the system will increase becomes virtually certain.

In any isolated system, however, entropy - or disorder - will keep increasing until the system reaches the state of maximum entropy known as "heat death" and all life ceases to exist. If, according to classical thermodynamics, the universe is moving towards such a state of maximum entropy and disorder, this tendential increase in disorder strongly contrasts, and even contradicts, the theory of evolution held by biologists, who believe that the living universe evolves from disorder to order and ever increasing complexity. This is, in nuce, the paradox of entropy.

Schrödinger's solves this paradox by stating the following: the organism escapes the decay to entropy through its metabolism, i.e., the exchange of energy and matter with its environment. He mentions activities - today endangered - such as eating, drinking and breathing as well as assimilation in the plant world. He recognizes the grim consequences of the law of entropy. If every process and event means an increase in entropy in the part of the world where it is taking place, then all organisms are doomed to approach the dangerous state of maximum entropy which leads to instability and destruction. But the history of life on earth shows us that this is not the case as species have indeed managed to survive. Schrödinger's lapidary answer to this paradox is: The essential thing in metabolism must be that the organism succeeds in freeing itself from the entropy which it cannot avoid producing while alive. The question that follows is how does the organism succeed in escaping thermodynamic equilibrium which means its sure death. Here Schrödinger introduces the notion of negative entropy. Negentropy is, firstly, a measure of order, but in his book it soon becomes something like a panacea against entropy. For example, when he states his bold hypothesis that living organisms feed on negative entropy.[5] An organism maintains its order and stability by extracting negative entropy from its environment. In spite of this brilliant general explanation Schrödinger could not provide convincing proof of how order emerges in nature from disorder or chaos. But he succeeded in giving research in this field a fruitful strategy for exploring how higher (aperiodic, living) order may have evolved from lower, periodic order, e.g., from crystals and periodic solids.

Nevertheless, Schrödinger's bold hypothesis of "higher order of life from primitive lifeless order" has been confirmed today by Onsager and Prigogine. (To illustrate this we will reconstruct Schrödinger's solution to the paradox of entropy in chapter 4, and compare it with Prigogine's solution.)

From the perspective of modern biology and chemistry, we have to be aware of the fact that an average cell is an enormously complex system, consisting of at least 1,000 different enzymes (each existing in 100 exemplars) i.e., 100,000 enzyme molecules, each of which perform a particular task, regulating the metabolism and controlling some 10,000 chemical reactions (syntheses) per minute. Such a complex, dynamic, self-replicating system surpasses anything that human technology has invented. The creative potential of one single cell can only be compared with all of the factories in this world, provided they were computerized, centralized, programmed, and fully automated. It is impossible to "reduce" such a system in a literal sense to simpler units without it losing its holistic order.

If we agree with Schrödinger that the cell's genome is order-sustaining and negentropic and that its environment is order-decreasing (entropic), we arrive at Schrödinger's paradox.[6] Today the structure of thermodynamics, as outlined by Schrödinger [7], can be broadly described in the following terms:

1) the conservation of energy
2) a steady increase in the entropy of its equilibrium states
3) any relaxation of the constraints on a system, which leads to an increase in its entropy. Schrödinger, who clearly saw this situation, so unfavorable for any living system, tried to explain the marvelous faculty of the living organism to maintain its order and to delay its decay to thermodynamic equilibrium or death.[8] We already know his answer: Living organisms feed upon "negative entropy" and maintain their order and organization by "extracting order" from the environment.[9] Schrödinger explains this in prophetic words: "Thus the device by which an organism is able to maintain itself stationary at a fairly high level of orderliness (= fairly low level of entropy) really consists in continually sucking orderliness from its environment." It is clear that his solution of the paradox only applies to the "order from order," or to order from the order-organizing genes, but not to order from disorder.

3. SCHRÖDINGER: THE INVENTOR OF OUR MODERN VIEW OF GENES - OR ORDER FROM ORDER

Schrödinger's radical proposal for cutting the Gordian knot, and solving the paradox of entropy, is to find the missing link between the negentropy (order) of living organisms and the entropy (disorder) of inorganic, environmental matter. This is only to be achieved by finding the hidden prestages of order in our entropic, inorganic world. But Schrödinger had trouble showing exactly how order may have emerged from disorder in the prebiontic world. To illustrate the transition from disorder to order, he cited only one example of a missing link: the order-preserving, inorganic, periodic crystal - a forerunner of the aperiodic solid - or the genome. Thus he assumed that even inanimate nature has an order-sustaining tendency which begins with periodic crystals, followed by aperiodic organic molecules. Schrödinger discovered that in complicated organic molecules, such as the genome, an additional order-sustaining property appeared, namely the storage of past experience. Every atom, and every group of atoms is packed with information, playing one specific role in preserving the order of the organism. It stays relatively invariant throughout the individual organism's life and throughout generations. With his missing link hypothesis which is based on the aperiodic crystal, Schrödinger explains how the complex order of living organisms evolved from the lower order of inorganic systems. " We believe a gene - or perhaps the whole chromosome fibre - to be an aperiodic solid."[10] Long before Crick and Watson, Schrödinger saw that genes are not only well-ordered string structures of atoms, but also the miniature code of life. Genes contain, according to Schrödinger, the code script ensuring the relative stability of all functions of the organism throughout their lifetime. They are not only the hereditary blueprints for replication, but also explain the unfolding of events in the life cycle of an organism, revealing admirable regularity and orderliness "unrivalled by anything we meet with in inanimate matter."[11] Genes control and concentrate a "stream of order" in the organism, enabling it to escape entropic decay into atomic chaos. For Schrödinger, genes represent the highest form of order we know of, much higher than the ordinary, periodic inorganic crystals.

According to Schrödinger, orderliness can be generated in two ways: 1) statistically, when order is produced from disorder in the inorganic field and, 2) biologically or genetically, in that higher forms of order are created from already existing lower order by means of the genes or Schrödinger's aperiodic solids. In the next chapter we want to concentrate on the first way of order generation.

4. ORDER FROM DISORDER: A HISTORICAL REVIEW - SCHRÖDINGER AND PRIGOGINE

Schrödinger's idea of how order is generated in prebiontic systems without violating the law of entropy, has been recently vindicated by the famous Onsager-Prigogine, dynamic model of thermodynamics.[12] Priogonine was the first thinker after Schrödinger to actually integrate the environment in thermodynamics. For this purpose he had to modify classical thermodynamics. Prigogine's basic assumption, already stated by Schrödinger, is that dynamic systems, which evolve over a period of time, and their environmental systems, form inseparable, interacting holistic units. This assumption was also the point of departure for Onsager's (1931) and Prigogine's modification (1962,1982) Systems are nothing without their environmental systems, and vice versa. Following Schrödinger, Onsager and Prigogine focused on the interaction and the complexity of the new holistic unit (the internal system and its environmental system) and were able to explain, as we will show later, why the internal system is able to maintain its order in spite of the law of entropy. The next important modification proposed by Prigogine was to transform thermodynamics into a non-linear theory. This change was prompted by the fact that not only living systems, but also inorganic systems undergo unforeseeable, temporal changes. In this case the maintenance of order in the internal system could be explained by invariance or stability conditions for the non-linear differential equations, which describe the temporal dynamics of the Onsager-Prigogine thermodynamic systems. Now, we are able to explain why lower, inorganic and lifeless dynamic systems which are embedded in their environments or niches, suddenly show astonishing, new order-maintaining properties. It seems as if Prigogine found the missing link between disorder and order in the prebiontic field, which Schrödinger had anticipated and expected. According to Prigogine even inorganic, dynamic systems can generate order spontaneously and maintain it, like higher-level living systems. Such systems resemble the most primitive, simple living systems, nude RNA-macromolecules, which Schrödinger's had called aperiodic solids. The recently discovered Tetrahymena[13] is such an example of a very primitive living organism.

The single-stranded RNA from the protozoan Tetrahymena Thermophila, which lives in pond water, catalyzes the cutting and splicing of its RNA that lead to the removal of a part of its own length and self-replication without the help of proteins or enzymes. Thus the Tetrahymena represents a form of precellular life which existed long before DNA, the cell, proteins or enzymes existed. Recent findings by Cech (1982,1986) have shown that protein synthesis, the most fundamental biosynthetic activity, can be catalyzed by RNA alone. That means that information-carrying capacity and catalytic activity are inherent in the same living molecule, i.e., RNA. Cech's discovery supports Eigen's and Schuster's idea that single-stranded RNA, rather than double-stranded DNA may have been the first and most important carriers of genetic information as life began on this planet. In other words, this newly discovered RNA can replicate itself "parthogenetically" without the help of proteins and enzymes and certainly is one of the oldest living organisms.

As we have already seen, lifeless inorganic, dynamic systems in Onsager's and Prigogine's non-linear thermodynamics have the most astonishing property, namely to spontaneously generate and sustain order. If the following subsidiary conditions hold, then this serves as new proof of Schrödinger's assumption that order can originate from disorder provided:

1) the dynamics of the inorganic, lifeless systems, their behavior, can be described in terms of the non-linear differential equations of Onsager's and Prigogine's thermodynamics.
2) the systems in question are open systems, and exchange matter and energy with the environment as living systems as our Tetrahemena RNA does
3) the system always remains far-removed from a state of an internal equilibrium - a condition Schrödinger had explicitly stated
4) autocatalytic and cross-catalytic reactions occur with feedback.

We know from classical thermodynamics and the paradox of entropy that any final state, any equilibrium state of lower inorganic systems must show a higher level of entropy than the initial one. But, such an increase in entropy would also increase disorder and uniformity and this would contradict the order-maintaining tendency of the living macromolecule Tetrahymena. It would disprove Schrödinger's order-from-disorder thesis. But if we apply Onsager's and Prigogine's subsidiary conditions, we are able to explain how and when order evolves from disorder. First, in modifying the linear classical thermodynamics to the non-linear one, we have to avoid the paradox by preserving, whether we want to or not, the most important, robust property of entropy increase. Second, we have to divide the total entropy S_t or the entropy change dS_t of the holistic system into two parts: $dS_t = DS + dS_e$, whereby S is the internal entropy of the (inorganic) system or of our Tetrahymena RNS, and S_e the entropy of the environment. Since the whole system S_t is an open system, it is evident that now, not even an internal-entropy decrease (equal to order increase) would be able to violate the second law of classical thermodynamics, if it is compensated by an equal increase of the external entropy S_e. This gives a precise illustration of Schrödinger's words: "living systems feed upon negentropy".

Thus, Prigogine and Onsager obtained the results Schrödinger had anticipated. Higher laws, for example, the order-preserving laws of the genes, can be imposed on lower, merely physical laws, since even lifeless, dynamic systems can become "negentropic" and order-sustaining, without violating the thermodynamic law of entropy.

The next question is: how can we circumvent the classical thermodynamic equilibrium? Generally speaking, each strict equilibrium would be the deadlock, the death of all evolutionary processes, the final stop, in Schrödinger's words. Again, Onsager and Prigogine's model will help us to avoid equilibrium states. It is a well-known fact that in inorganic systems, not far from the internal equilibrium, the changes or deviations from the equilibrium become smaller and smaller, or the gradient (the rate of changes with respect to the distance) approaches zero.

Now, all changes within our dynamic, holistic system S_t (internal system plus environmental system) consist of flows F of energy and heat (expressed by their gradients G). Since each flow F_i is mutually dependent on all other flows F_j, we obtain the total flows $F_{ij} = F_{ij}x_ix_j$. According to Onsager we get some kind of invariantly stable state of the dynamic, internal process, if dG/dt is a minimum or the internal entropy production approaches a minimum, irrespective of the sum total of the entropy of the whole system. Therefore, even inorganic systems may strive towards a relatively stable state by increasing their order, in short, they behave like living systems.

These new findings obviously confirm Schrödinger's thesis ("order of living systems from inorganic order and inorganic order from inorganic disorder"). We have evidence that there is already a gradual increase in order and negentropy within the inorganic dynamic and lifeless systems without violating the law of entropy. The process in which order emerges from disorder starts already at the lifeless inorganic level and is a prestage of the biological self-organization at a higher level. Here we have the missing link in the emergence of order. In other words, biological evolution, as the production of evolutionarily stable states of living systems, presupposes the generation of relatively, or dynamically stable states of inorganic systems. Dynamically or evolutionarily stable means, to quote Prigogine, "near the equilibrium states".

In greater detail, inorganic systems are always disturbed by molecular fluctuations, Brownian molecular movements and random events from the environment. Therefore any temporarily or relatively stable, steady state will be sooner or later disturbed by random events. If such disturbances lead to symmetrical oscillations, the system will remain near equilibrium or relatively stable. But, if it is disturbed in an asymmetrical sense, it will leave its "near the equilibrium $state_1$" at time t_1, in favor of a new "near the equilibrium $state_2$" at time t_2. It all depends on two further conditions:

1) if the asymmetric disturbances or fluctuations disappear, the inorganic systems remains stable, near equilibrium $state_1$
2) if the asymmetric disturbing fluctuation prevails, it automatically changes over to a new $state_2$ close to equilibrium. This explains how relatively stable system states develop. If a system suddenly becomes unstable, it will change from equilibrium $state_1$ to the new equilibrium $state_2$. If now the new $state_2$ close to equilibrium produces less entropy, its internal order (organization) is, of course, greater at the cost of its environment.

A higher-level inorganic state can be generated from a lower-level inorganic state of lower order or even disorder, provided that

1) the higher ordered system is open, i.e., reacts like a living system to its environment
2) the higher-level system is always far from the internal deadlock of a stationary equilibrium
3) it possess auto- and cross-catalytic reactions, which again are necessary to create negentropy.

Such systems very soon behave like biological systems fighting against mutants (random events) in small numbers equivalent to symmetrical disturbances or against mutants in great numbers (asymmetric disturbances). The more stable states or the better adapted will "survive." The new non-linear version of Prigogine's thermodynamics permits, of course, the interaction between lifeless systems as well as living systems and their environments or niches, such as our Tetrahymena RNA.

There is one fundamental difference between order-sustaining lifeless, dynamic systems and living systems. In spite of the fact that Prigogine's thermodynamics circumvents entropic disorder, it only guarantees short-living, non-inheritable maintenance of order. Only the higher laws of the genome guarantee the inheritance and maintenance of order, since the life-sustaining experience of earlier generations is contained in the genes in codified form.[14]

All of this has some interesting, philosophical consequences.

1) We cannot reduce living systems to inorganic, physical levels. Even if we assume that living systems are stratified and composed of living and inorganic levels, reduction would eliminate the influence of higher laws on a physical level. The higher laws can only apply within a living system. Only within these systems are they superimposed on their inorganic levels, forming one holistic, interacting system. Therefore lower levels cannot replace higher levels, and vice versa.[15]
2) Ontology, no longer a static, traditional discipline which describes neatly separated and hierarchically ordered levels, is now a dynamic, evolutionary ontology of becoming. Its purpose is to explain the increase of relatively stable order, and stability of inorganic and living systems during the course of evolution. This new ontology has to take into account that the traditional barriers between different levels simply break down in all living systems.[16]
3) The RNA of our Tetrahymena and the DNA of higher organism stay relatively stable. They "always remain the same," as Schrödinger stated. In other words, over the millenia they have built up an interindividual memory which contains rules regulating not only the metabolism and the replication of the organisms, but also the relative maintenance of the physical, physiological, psychic and mental order throughout the life of any organism.

5. SCHRÖDINGER THE PHILOSOPHER: THE PHILOSOPHY OF THE 'I' AND THE GENES

Already in 1944 Schrödinger foresaw that our view of man and the world would be revolutionized by the integration of quantum physics, quantum chemistry, biology and genetics. He was convinced that future research would reveal that the higher laws of life are codified in our genes. He began to draw astonishing philosophical and social consequences from the new results of molecular biology and genetics. In fact, he constructed a general philosophical framework for quantum theory, biology, genetics and philosophy which is basically a new "philosophy of the genes." In "On Determinism and Free Will", the last chapter of his book *What is Life?*, Schrödinger introduced new, alternative philosophical views of various traditional issues, ranging from a new definition of the genome through philosophy and religion to a reevaluation of the question of personality and the self. Here Schrödinger was no longer interested in the genome as the basic structure of heredity, but in the genome as the carrier and guarantor of the individual's invariant and characteristic personality.

When Schrödinger compared the relative stability of our 'I' during our life with the relative stability of the genes, he concluded that the stability of our personality or the 'I' of a normal person is indirectly or directly due to the stability of our genes. Western and especially Eastern (Indian) philosophers and religions have always equated our personality and the 'I' with our soul, and regarded our soul as something in us which always remains the same. The importance of the genome, as Schrödinger saw it is that "it is the miniature code" and corresponds directly to a specific, highly complicated scheme according to which our characteristic personality, or our 'I' is developed and maintained. [17]

The most important aspects of his philosophy of the genes are as follows:

1) The normal and healthy individual's genes remain the same throughout his or her life .
2) Since the individual's genes maintain bodily and psychic order during his or her lifetime, the human genome is responsible for maintaining the mental order of our individuality. Thus, genes guarantee that we maintain the same personality, which some call self-identity and some call soul.[18]
3) In Western philosophy, the soul has been regarded as the permanent or substantial form or order of the body, and the only source of all vital and mental performances since Aristotle. There are Western philosophers and Christian thinkers who see the human soul as a spiritual, non-empirical and immortal substance. By contrast, empirical and Eastern philosophers, Schrödinger and Eastern religions regard the soul as empirical, finite and unchangeable only for the duration of our life time. In both views the soul, our personality, or our 'I' forms, orders and unifies in a characteristic and unique way the fleeting succession of our conscious experience, creating and regulating our characteristic behavior.

The development of genetics since Schrödinger, especially the dramatic turn in the exploration of the genes brought on by Crick and Watson in the fifties, has given way to a similar "empiristic" and finite philosophical view. There is growing evidence for the dependency of our personality, our 'I', or soul on our genes which confirms Schrödinger's philosophy of the genes and the I.

To sum up these arguments: today the genes are regarded as the determining factors of the species, of individual life and of our biological evolution. The importance of the genes for the species was obvious when biologists began to decipher and read the information stored in the genes of our chromosomes. Suddenly genes turned out to be a DNA or nucleotidian memory of the species, where experiences from the past of a species were stored. Genes served as the ultimate physical carriers of information, storing and transmitting to following generations, inheritable characteristics optimal for the species and the individual's survival.

Surprisingly enough, geneticists and biologists have found that genes are even more important for the individual's personality than is normally assumed. In the genes we find - as Schrödinger anticipated - the blueprints, not only for the construction of our individual body, for our metabolism and our individual sexual behavior, but also instructions (rules) for our psychic and mental and, according to Lumsden and Wilson, even for our characteristic socio-cultural and individual behavior.

Yet even more astonishing is the fact that they regulate our characteristic, individual social behavior and customs, such as: egotistic and altruistic feelings, courtship, discrimination between ingroups (family, hordes, groups) and outgroups, to mention only a few examples. In short, our individual personality and characteristics and our socio-cultural behavior were supposed to be preprogrammed and regulated by the genes.

Thus, genes regulate and are partly responsible for not only the hereditary mechanisms and the struggle for survival, but exactly for what Schrödinger called the 'I'. According to his philosophical view, it is the 'I' which stands for the same physical, psychic and mental order and for the relative stability of our individual, distinctive anatomical, brain physiological, cognitive and behavioral traits of our organism for as long as we live. The genes are thus responsible in an empirical sense for the specific way an individual feels, follows inclinations, evaluates, behaves socially, imagines things and thinks.

Genes, however, are not stupid, political dictators. They reward, as Mach and Schlick had anticipated, in an "intelligent" and clever way, certain genetically optimal actions, customs and, last not least, sex and social enjoyments with a plus of pleasure or pleasant feelings; and they suppress unfavorable ones through the accompanying displeasure. Fortunately, they can make us inclined to do something, but they cannot force us, which is exactly what Schrödinger called statistical determinism.

From his conception of the stability of the human genes (the genome) Schrödinger derived the stability of the 'I,' or what we call our soul in our religious, Western tradition. Our soul exists in us permanently but only for the duration of our lifetime. It determines our characteristic personality. But, contrary to the conception of an eternal soul, Schrödinger endows the soul or what he calls the 'I' with only a finite lifetime. Thus the 'I', or our soul, lasts only as long as we exist.

Since only the genes in our body remain stable throughout our lifetime, only they can guarantee our self-identity. Therefore, it is clear why Schrödinger based the 'I,' the inner order of each individual's personality on the empirical order of our genes. This conclusion let him discover further, original philosophical implications regarding consciousness, the finiteness of the individual soul, the individual self and the 'I'. These implications turned out - as he himself underlined - to be alien to our Western and Christian ideology about the soul, but can be found in Schopenhauer as well as in Eastern and Indian philosophy and religion.

Strangely enough, it was the rational physicist Schrödinger who triggered the exodus of the "discontents" from our Western culture and from our overly traditional Western ideology. Since Schrödinger there is the lingering suspicion that the world view imparted by modern biology and physics is - at least partly- inconsistent with our traditional, Christian culture (but not with our Christian ethics).

Schrödinger, who regarded himself as "statistico-deterministic" is in modern terms a "weak interactionalist". This means, that body and mind only interact in a statistical, causal sense.[19] He rejects rigid determinism for two reasons. One, if our body's functions determine in a deterministic sense, our feelings, imaginations and our mind, then our free will would be impossible. Two, free will does not mean: I influence my motions, my bodily functions. If I am the person, who controls even the motion of the atoms in my body, then a biologist would get easily entangled into the blasphemous and lunatic inference "Hence I am God Almighty."[20]

Yet - according to Schrödinger - neither our mind nor our consciousness (i.e., subjective awareness or self-awareness) controls and superimposes our will on our body's reactions. Rather, it is something far more complicated, namely the interaction between the genes and the 'I', in particular the mutual interrelations between the genes and the 'I'. This interaction between the 'I' and the genes resembles Freud's and Jung's interrelation between the conscious and the unconscious. It is a strange coincidence that Freud always assumed that future research would find the ultimate physical and chemical basis of the unconscious. Freud saw the unconscious as the source of the individual's characteristic, conscious behavior and for Jung the unconscious consisted of behavioral archetypes, specific instructions hidden away in the unconscious which surface on the conscious level, telling us how we should behave or solve a conflict. The psychoanalytical unconscious stays - like the genes - relatively stable throughout our lifetime. For Freud, conscious awareness represents only a thin upper layer resting on the hidden regions of the instinctual forces of the unconscious. Jung divided the unconscious into the personal and the collective unconscious, in a similar way biologists distinguish between individual genome and the gene-pool. The personal unconscious contains specific, individual patterns of behavior, of feeling and thoughts that are subject to conscious recall. By contrast, the collective unconscious is an assemblage of the life experience of past generations which regulates and influences the psychic functions, ideas, modes of thoughts, patterns of reactions common to all humans.

It is important to note that Schrödinger used 'I' only for the single, unique individual. The 'I' begins to take shape in our youth when the individual's unique character and personality is in the process of being formed and specialized. It is influenced in particular by the education and the development of the individual. Schrödinger, the German idealistic philosophers, Schopenhauer and Freud, never used the conceptualized form of 'I' (which in English is "ego"). It was a big mistake to translate Freud's 'I' with the English word "ego". Only the word 'I' reveals what he and Schrödinger really meant, namely the concrete, unique and empirical person. One does not say "ego go to the shop" but "I go to the shop".

Introspection teaches us that my 'I'[21] or consciousness, or my mind is never experienced in the plural, only in the singular. Similarly, from genetics we have learned that each genome differs from the other. Even split consciousness, double personality, or several characters appearing in one dream are never manifested simultaneously. Consciousness is always intimately connected with, and dependent on, the physical state of the individual body. Therefore we should become suspicious of the plurality hypothesis, which assumes the existence of billions of immortal souls existing by themselves or of collective souls.

As Schrödinger puts it: "The only possible alternative is simply to keep the immediate experience that consciousness is a singular of which the plural is unknown."(*What is Life?*, p. 87). That there are innumerable souls of the billions of humans which ever existed, is a myth produced by wishful thinking or deception (the Indian MAJA).[22]

Each of us has the distinct impression that the total sum of his or her own characteristic experiences and memories form a unique non-repeatable unity, quite different from that of any other person. Everybody refers to it as my 'I'. It is more than a collection of past experiences and memories. Memories are like the canvas upon which the pictures of a painter or a movie appear. But on close introspection, what is really meant by 'I', is that what Schrödinger called the "basic material," the painter, the movie projector, the genes. Like Schrödinger, I, too, left my native country and spent 20 years in the United States. I can deeply understand what Schrödinger wrote at the end of his book: "You may come to a distant country, lose sight of all your friends, may all but forget them; you acquire new friends, you share life with them as intensely as you ever did with your old ones." (*What is Life?*, p. 88). The 'I', or the person stays the same, creating again and again the same way of living. But memory of the past is not everything. Schrödinger continues: "And even if a skilled hypnotist succeeded in blotting out entirely all your earlier reminiscences, you would not find that he had killed you. In no case is there a loss of personal existence to deplore. Nor will there ever be."

With the help of our brain, the genome reconstructs our world, allowing us to feel identical with the "youth that was I" This 'I' is the experience of an inner stability of one's self identity which accompanies us like a guardian angel as long as we live. The reason for this is, according to Schrödinger, that our memories, our 'I', hinges upon a solid, the aperiodic crystal, as he described the genes, the "finest masterpiece ever achieved along the lines of the Lord's quantum mechanics."[23]

NOTES

1 Schrödinger, Erwin, *What is Life?*, Cambridge University Press, Cambridge 1948, p. 76.
2 Prigogine, I., *Vom Sein zum Werden*, Zurich 1982, p. 249.
3 Smith, John Maynard, *Evolution and the Theory of Games*, Cambridge University Press, New York 1982, p. 11-26.
Leinfellner, Werner, "Evolutionary Causality, Theory of Games, and Evolution of Intelligence," in: Wuketits, M. (ed.) *Concepts and Approaches in Evolutionary Epistemology*, Reidel, Boston 1984, p. 261.
Leinfellner, Werner, "Evolutionäre Erkenntnistheorie und Spieltheorie," in: Riedl, R., Wuketits M. (eds.), *Die Evolutionäre Erkenntnistheorie*, Parey, Hamburg 1986, pp. 195-210.
4 Schrödinger, E., *What is Life*?, loc. cit., p. 68.
5 ibid., pp. 69-75.
6 ibid., p. 79.
7 ibid., pp. 68-75.
8 ibid., p. 74.
9 ibid., p. 74.
10 ibid., p. 61.
11 ibid., p. 77.
12 Prigogine, I., *Non-Equilibrium Thermodynamics*, New York 1962.
Prigogine, I., *From Being to Becoming*, 1982, see note 2.
13 Cech, T. R., "Self-Splicing RNA: Autoexcision and Autocyclization of the Ribosomal RNA. Intervening Sequence of Tetrahymena," in: *Cell*, vol. 31, no. 1, 1982, pp. 147-157.
Cech, T.R. et al., "The Intervening Sequence RNA of Tetrahymena is an Encyme", in: *Science*, vol. 231, no. 4737, 1986, pp. 470-475.
14 Smith, John Maynard, *Evolution and the Theory of Games*, Alden Press, Oxford 1982, pp. 13-17.
Concepts and Approaches in Evolutionary Epistemology, Reidel, Boston, pp. 233-275, in particular, p. 261, Leinfellner, W.
15 Leinfellner, Werner, "The Change of the Concept of Reduction in Biology and in the Social Sciences", in: Radnitzky, G. (ed.), *Centripetal Forces in the Sciences*, Paragom House, New York 1988, pp. 55-79.
16 Leinfellner, W. and E., *Ontologie, Systemtheorie und Semantik*, Duncker & Humblot, Berlin 1978.
Leinfellner, W., "Grundtypen der Ontologie," in: R. Haller (ed.), *Language, Logic and Philosophy*, Hölder-Pichler Tempsky, Vienna, pp. 124-131.
17 Schrödinger, E., *What is Life?*, Cambridge University Press, Cambridge 1948, p. 68.
18 ibid., p. 88 ff.
19 Leinfellner, W., op. cit. 1984, pp. 237-245.
20 Schrödinger, op. cit., p. 89.
21 ibid., p. 91.
22 ibid., p. 91.
23 ibid., p. 86.

ERWIN SCHRÖDINGER AND HIS CONTRIBUTION TO A NEW UNDERSTANDING OF LIVING SYSTEMS

FRANZ M. WUKETITS
Konrad Lorenz Institut für Evolution und Kognition
Adolf Lorenz Gasse 2, 3422 Altenberg, Austria

1. INTRODUCTION

In the great Encyclopedia Britannica one finds Erwin Schrödinger portrayed as an outstanding scholar, a scientist displaying extraordinary intellectual versatility:

> Of all the physicists of his generation Schrödinger stands out because of his extraordinary intellectual versatility. He was at home in the philosophy and literature of all of the Western languages, and his popular scientific writing in English, which he had learned as a child, is among the best of its kind ... Because of his exceptional gifts Schrödinger was able in the course of his life to make significant contributions to nearly all branches of science and philosophy, an almost unique accomplishment at a time when the trend was toward increasing technical specialization in these disciplines.[1]

Surprisingly enough, when I studied biology at the University of Vienna, none of my teachers - except for Rupert Riedl - mentioned Schrödinger and his ideas. Riedl, however, pointed to Schrödinger's important contribution to the life sciences and explained his (Schrödinger's) concept of *negative entropy*. This concept has indeed been extremely important in biology, because it helps us to understand the physical basis of order in living systems.[2] Thus, Schrödinger is one of the important figures in the development of biology in the twentieth century.

In the present paper I shall review the meaning of Schrödinger's work in the field of biological science. This paper is divided into four parts: first, I want to touch on Schrödinger's approach towards a synthesis in science; second, I shall show how Schrödinger's work forms a cornerstone in the development of philosophical issues in the life sciences; third, I shall recall his significant contribution to molecular biology, keeping in mind that he was one of the pioneers, if not the founder, of this discipline; fourth, I want to show that Schrödinger has paved the way to some comprehensive ideas in today's science.

2. THE NEED FOR SYNTHESIS

Modern science is characterized by a trend toward increasing specialization. This trend began in the last century and has found its clearest expression during the last three decades or so. Schrödinger realized the need for a synthesis in science. As he said:

> We have inhrerited from our forefathers the keen longing for unified all-embracing knowledge. The very name given to the highest institutions of learning reminds us, that from antiquity on and throughout many centuries the universal aspect has been the only one to be given full credit.[3]

J. Götschl (ed.), Erwin Schrödinger's World View, 99–105.

But at the same time Schrödinger realized the enormous specialization in science, the "atomization of knowledge", as it were. Thus he wrote:

> But the diffusion, both in width and depth, of the multifarious branches of knowledge during the last hundred odd years has confronted us with a strange dilemma. We feel clearly that we are only now beginning to acquire reliable material for welding together the sum total of all that is known into a whole.[4]

Schrödinger formulated an important desideratum which is today perhaps even more important than some forty years ago. However, he also saw the difficulties of establishing a synthesis. "It has become", he said, "next to impossible for a single mind to fully command more than a small specialized portion of it"[5] (i.e., of the whole corpus of knowledge). So how was one to overcome this dilemma?

> I can see no other way out (lest our true aim be lost forever) than for some of us to work toward a synthesis of facts and theories, albeit with second-hand and incomplete knowledge of some of them - and at the risk of making fools of ourselves.[6]

Schrödinger was one of those scholars who sought a synthesis - without making a fool out of himself!

Our world view has indeed become fragmented. Philosophy, for example, has broken away from the (natural) sciences. It should be clear, however, that there is no science without philosophical questions and that, by the same token, the philosopher needs the sciences if he or she does not want to build castles in the air. It is obvious that Schrödinger was a "philosophical person", as can be seen in his *Weltansicht* [7] and from his discussion of the question *What is Life?* (c.f. the following paragraph.)

Here I want to sketch Schrödinger's attempt to link physics with biology. One of the greatest challenges here is to link the properties of living systems with physical principles without reducing the former to the latter. Many people have learned to see physics as deterministic and mechanistic and biology as the study of systems displaying an enormous versatility. On the one hand, there is a fundamental law according to which everything in this world tends towards increasing thermodynamical equilibrium (entropy) and on the other, there is order and complexity, organisms as systems far from thermodynamical equilibrium and thus producing negative entropy. On the one hand, we have dead matter, and on the other living beings as developing, self-reproducing systems with their own history. This state of affairs is anything but ideal. But perhaps the notion of physics underlying such a view is incorrect.

From the perspective of physics the universe is in a way "ill-disposed" towards life;[8] the emergence of life on earth was not highly probable, and the reader may recall Monod's theory according to which the origins of life was an accidental event.[9] "Life," as Grobstein puts it, "constitutes a spreading center of order in a less ordered universe."[10] This is a paradox: On the one hand, the universe tends towards increasing chaos, on the other there is a small, a very small part in this universe which contains living systems, i.e., systems of increasing order and complexity. However, this only appears to be a paradox, but actually - as Schrödinger showed - it is not a paradox, because it follows from physical principles.

Take Schrödinger's concept of negative entropy. This concept has taught us that organisms do not contradict the principle of entropy but that they *-as open systems-* are able to "circumvent" it.[11] Hence, any living system is able to arrange order. Already Ludwig Boltzmann proposed a statistical, mechanical definition of entropy:

$$S = \underline{k} \log W + C$$

Schrödinger used well-defined concepts of "order" and "disorder" to arrive at a clear definition of negative entropy. If, he argued, D is a symbol for disorder, then 1/D may be used to understand the dimensions of order in a (living) system. He then attained the following formula:

$$\underline{k} \log (1/D)$$

which is the formula for negative entropy. Schrödinger explained that "the awkward expression *negative entropy*" can be replaced by a better one: entropy, taken with the negative sign, which serves as a measure of order."[12]
He then continued:

> Thus the device by which an organism remains stationary at a fairly high level of orderliness (= fairly low level of entropy) really consists in continually sucking orderliness from its environment. This conclusion is less paradoxical than it appears at first sight.[13]

Hence Schrödinger's arguments are not only clearly presented but also rather simple. As Yourgrau comments: "His reasoning is extremely simple and convincing and expressed in such a manner that one is prone to forget that the gist of his derivation can already be found in Boltzmann's celebrated entropy function."[14]

However, what Schrödinger established was a physical principle underlying living systems. Hence, he contributed much to a synthesis of physical and biological science: He demonstrated that there is no gap between living and non-living matter. In this respect his work is also of great philosophical import.

This leads us to an old and venerable philosophical enigma.

3. WHAT IS LIFE?

Schrödinger discussed this question from the perspective of the physicist, without denying that living systems are the most unique of all physical systems. Scientists often disregard this and argue that it has not so much to do with science than with metaphysics. This is incorrect. The question "what is life?" is not necessarily a metaphysical one: Schrödinger was one of the scholars of the twentieth century who showed that the question is indeed amenable to science.

The situation in the life sciences at the beginning of the twentieth century was hardly a satisfactory one. There were two different interpretations of life, two different "philosophies" of biology: vitalism versus mechanism. The vitalists claim that life is directed by particular vital forces or spritual principles; mechanism, by contrast, claims that living systems can be reduced to mechanical principles.[15] There was, in other words, a materialistic and an idealistic biology and biophilosophy. What made these two positions seem so inadequate? The vitalists reduced life to "something" which remains unexplained or even unexplainable, whereas the mechanists simplified life and its particular properties by arguing that a living system is nothing other than a mechanical object. Both failed to see the very nature of life.[16] On the one hand, life was metaphysically obscured in vitalism, on the other hand, the essential characteristics of living systems were not adequately explained in mechanism.

Schrödinger realized that living beings, with their order and complexity, differ from non-living matter (and from the laws of classical mechanics), and that living order is able to reproduce itself, i.e., to create, again and again, highly organized systems, systems of order.[17] He argued that living systems operate in a manner which cannot be reduced to the laws of classical physics (mechanics). But he also argued that this irreducibility of life cannot be properly explained by taking recourse to some "new force". The construction of a living system itself would be the key to an understanding of the nature of life:

> What I wish to make clear...is, in short, that from all we have learnt about the structure of living matter, we must be prepared to find it working in a manner that cannot be reduced to the ordinary laws of physics. And that not on the ground that there is any 'new force' or what not, directing the behaviour of the single atoms within a living organism, but because the construction is different from anything we have yet tested in the physical laboratory.[18]

I think that from the perspective of today's biology, Schrödinger's view holds true.

But let us direct our attention to a more general aspect. In dealing with the question "what is life?" Schrödinger continued the discussion of an old philosophical issue. As I said before, this question is not necessarily a metaphysical one. But it does reflect important philosophical problems. Moreover, the question is of some importance to cultural philosophy: our answer is decisive for the way we look at life in a concrete sense. If I say "life is nothing else but mechanics", I might feel prompted to treat living systems as mechanical systems, that is to say, brutally. The theoretical answer to the question can have - and actually has had - practical consequences. Thus, living systems must be seen as physical objects - and not as "spirtual beings" or the like -, but at the same time we must be aware of the fact that they are most unique physical objects with a specific meaning and value for us.[19]

Schrödinger was aware of the philosophical implications of the question "what is life?" and tried to deal with this question in a most comprehensive manner. He was concerned with four particular questions:[20]

(i) How an organism can manage to escape from the tendency to destroy life organization, i.e., how living matter evades the decay to equilibrium?
(ii) What is the nature of the hereditary substance?
(iii) How does this substance reproduce itself?
(iv) What is the nature of consciousness and free will?

I shall refrain here from going into great detail. But, with regard to the first question, it may suffice to remember Schrödinger's principle of negative entropy. The second and third question have been answered by molecular biology (cf. the following paragraph); the fourth question clearly shows Schrödinger as a "philosophical man", dealing with the great problems of the human intellect.[21] Undoubtedly, Schrödinger tried to see scientific problems in a broader context, that is to say, in the context of cultural history and the history of philosophy.

4. SCHRÖDINGER AND THE BEGINNINGS OF MOLECULAR BIOLOGY

Schrödinger is one of the important figures in the history of biology in this century. He was one of the pioneers of molecular biology and is one of the leading representatives of the "romantical phase" in the development of this discipline,[22] a phase characterized by great enthusiasm and hopes that the nature of heredity could be explained in terms of physics. Allen comments on the situation as follows:

As one of the founders of quantum theory, Schrödinger's shift of interest to biology was significant and caused a number of physicists to take notice. Attracted by the possibility of finding "other laws of physics", Schrödinger's approach led some physicists into what appeared to be the new frontier of scientific investigation. It does not take a psychological analysis to see how such an approach could attract a number of inventive, even romantically inclined physicists who had come to feel that the only work remaining in their field was that of clarifying details.[23]

Here is not the place to tell the story of genetics and molecular biology.[24] I shall only make a few brief remarks as far as this story is linked with Schrödinger's work. Once more arguing that living systems are unique in some way or other, Schrödinger said:

The unfolding of events in the life cycle of an organism exhibits admirable regularity and orderliness, unrivalled by anything we come across in inanimate matter. We find it controlled by a supremely well-ordered group of atoms, which represent only a very small fraction of the sum total in every cell. Moreover, from the view we have formed of the mechanism of mutation we conclude that the dislocation of just a few atoms within the group of "governing atoms" of the germ cell suffices to bring about a well-defined change in the large-scale hereditary characteristics of the organism.[25]

Schrödinger distinguished between two mechanisms which cause orderly processes in the organism: a statistical mechanism producing order based on disorder, and another mechanism, unique to living systems, producing order based on order.[26] And with regard to the particular mechanism that maintains an organism's degree of order, he, again, referred to the principle of negative entropy.

Schrödinger's view of hereditary principles is based on the hypothesis that a small molecule, the "germ of a solid" is the starting point for constructing increasingly larger relationships. In the living world, Schrödinger argued, this means;

that of building up a more and more extended aggregate without the dull device of repetition. That is the case of more and more complicated organic molecule in which every atom, and every group of atoms, plays an individual role, not entirely equivalent to that of many others (as is the case in a periodic structure). We might quite properly call that an *aperiodic crystal* or solid and express our hypothesis by saying: We believe a gene...to be an aperiodic solid.[27]

This view is based on the old analogy, according to which an organism functions similar to clocks. With his ideas, Schrödinger paved the way to new approaches in empirical (molecular-biological) research as well as to new theoretical conceptions, his most important step in this context being the combination of the definition of the hereditary substance (= aperiodic crystal) with the theory of chemical compounds and the theory of genes and mutations.[28] Once more it becomes clear that Schrödinger realized the need for a synthesis in the sciences.

Generally speaking, it can be said that Schrödinger provided the theoretical foundations for molecular biology. His ideas helped molecular biologists to pursue research and to finally attain a model of the molecular code of life, i.e., the structure of the DNA.[29] The recent development of molecular biology has confirmed the main lines of Schrödinger's argumentation and already in the fifties it became evident that the major expressions of life are based on the structure of organic molecules [30] although this does not mean that all functions and behavioral patterns proper to living systems are nothing else but molecular, so that a holistic, systems-theoretical approach is needed.[31]

5. SCHRÖDINGER AND CURRENT ISSUES IN SCIENCE

Undoubtedly, Schrödinger's work is most important in view of current developments in science, both physics and biology. Whenever a synthesis in science has been attained, there have been important innovations. With regard to Schrödinger's work - most notably *What is Life?*[32] - the synthesis of physical and biological concepts has been useful in a heuristical sense. He intended to bridge the old gap between inorganic and organic matter.

Today the conception of self-organization of matter has assumed greater importance. This conception implies that there is a tendency of particular types of matter to create self-reproducing systems.[33] Thus, life, its emergence some billions of years ago, is to be explained by some general properties of matter. But also our view of matter has changed: matter is not just something dead, it is rather a dynamical principle.[34] Life, in that sense, is a "dynamical state of order" of matter.[35]

Generally, we see the universe as a gigantic, self-organizing system, and self-organization and development as fundamental categories of our understanding of the cosmos. But this also evokes older ideas: *panta rhei*.

The implications of a "universal theory of evolution" - which, of course, is not something altogether new - should become fully recognized.[36]

Schrödinger made a decisive step towards such a theory, and his general (philosophical) reflections are still up-to-date. They are in tune with current ideas in science, as, for instance, the concept of self-organization.

To be sure, the development of modern science is characterized by extremely rapid progress. For this reason, we should always keep in mind the more general aspects - in particular, synthesis, which has become more and more important in science. Schrödinger was one of the people who recognized the need for synthesis decades ago. It might be that some of his philosophical conceptions were rather vague. But it is true that "if a man never contradicts himself, the reason must be that he virtually never says anything at all."[37]

Schrödinger said many things - and what he said is worth exploring.[38]

NOTES

1 *The New Encyclopedia Britannica*, vol. 10, Encyclopedia Britannica Inc., Chicago 1985, p. 538.

2 See R. Riedl, "*Order in Living Organisms: System Conditions in Macro-Evolution*, Wiley, New York 1979. (See also R. Riedl, in the present volume.)

3 E. Schrödinger, *What is Life*? & *Mind and Matter*, Cambridge University Press, London 1967, p. 1, (*What is Life*? is based on lectures which Schrödinger had delivered in 1943; *Mind and Matter* originated from lectures delivered in 1956.)

4 ibid., p. 1

5 ibid., p. 1

6 ibid., p. 1

7 E. Schrödinger, *Meine Weltansicht*, Zsolnay, Vienna-Hamburg 1961.

8 See on this F. M. Wuketits, *Zustand und Bewußtsein: Leben als biophilosophische Synthese*, Hoffmann und Campe, Hamburg, 1985.

9 See J. Monod, *Chance and Necessity*, Knopf, New York 1971.
10 C. Grobstein, *The Strategy of Life*, Freeman, San Francisco 1974, p. 7.
11 On the notion of open systems see particularly the works by Ludwig v. Bertalanffy, e.g., his *General System Theory*, Penguin Books, Harmondsworth 1973. On the relevance of this notion in contemporary biology see also F. M. Wuketits, *Biologische Erkenntnis: Grundlagen und Probleme*, Fischer, Stuttgart 1983.
12 E. Schrödinger, op. cit., note 3, p. 79.
13 ibid., p. 79.
14 W. Yourgrau, "Marginal Notes on Schrödinger", in A. D. Breck and W. Yourgrau (ed.), *Biology, History and Natural Philosophy*, Plenum Press, New York, p. 340.
15 Full details in F. M. Wuketits, op. cit., note 8.
16 ibid.
17 E. Schrödinger, op. cit., note 3.
18 ibid., p. 81.
19 Hence, there are connections between biology, natural philosophy, and ethics. See F. M. Wuketits, op. cit., note 8.
20 See E. Schrödinger, op. cit., note 3. See also R. Olby, "Schrödinger's Problem: What is Life?", *Journal of the History of Biology* 4, 1971, pp. 119-148.
21 See on this his *Mind and Matter*, note 3.
22 See G. Stent, "That was the Molecular Biology that was", Science 160, 1968 (pp. 390-395). See also E. Oeser, *System, Klassifikation, Evolution*, Braumüller, Vienna-Stuttgart 1974.
23 G. E. Allen, *Life Science in the Twentieth Century*, Cambridge University Press, London 1978, p. 198.
24 See on this J. D. Watson, *Molecular Biology of the Gene*, W. A. Benjamin, New York-Amsterdam 1965.
25 E. Schrödinger, op. cit., note 3, p. 82.
26 ibid.
27 ibid., p. 65.
28 On genes from the viewpoint of molecular biology see again J. D. Watson, op. cit., note 24.
29 The story of the discovery of the DNA and its structure is told by J. D. Watson, *The Double Helix: A Personal Account of the Discovery of the Structure of the DNA*, New American Library, New York-Toronto 1968.
30 See W. Wieser, *Gewebe des Lebens: Ein Leitfaden durch die zeitgenössische Biologie*, Schünemann, Berlin 1959.
31 See P. Hoyningen-Huene and F. M. Wuketits (eds.), *Reductionism and Systems Theory in the Life Sciences*, D. Reidel, Dordrecht-Boston-Lancaster 1989.
32 See note 3.
33 See M. Eigen and R. Winkler, *Das Spiel: Naturgesetze steuern den Zufall*, Piper, Munich - Zurich 1975.
34 See F. M. Wuketits, op. cit., note 8.
35 See M. Eigen, *Stufen zum Leben: Die frühe Evolution im Visier der Molekularbiologie*, Munich-Zurich 1987.
36 A recent attempt to develop such a universal theory of evolution and to show philosophical implications see H. Hörz and K.- F. Wessel, *Philosophische Entwicklungstheorie*, VEB Deutscher Verlag der Wissenschaften, Berlin 1983.
37 Miguel de Unamuno, quoted by Schrödinger, op. cit., note 3, p. 81.
38 This paper is a slightly modified version of a paper published also in German.

PART III

ERWIN SCHRÖDINGER'S WORLD VIEW AND HIS PHILOSOPHICAL SYSTEM

ON THE SCIENTIFIC AND HUMANISTIC THINKING OF ERWIN SCHRÖDINGER DURING HIS LAST YEARS

LEOPOLD HALPERN
Department of Physics, Florida State University
Tallahassee, Florida 32306, USA

In this day and age there is hardly any natural scientist who reflected on our existence from such a great variety of perspectives as Erwin Schrödinger did.

As a young man he had aspirations of becoming a poet and philosopher, but for financial reasons he was only able to study physics, mathematics and philosophy. He was, however, always firmly convinced that the true purpose of science is to illuminate the mystery of our being and to provide us with a better understanding of who we are.[1]

The broadest contemplation of all phenomena combined with the most profound intellectual analysis appeared to him to be the best way to achieve this goal. No phenomenon was to be left unconsidered and every aspect of our intellectual creations studied so that we may be able to understand who we are. This even applies to theology (which often enough has declared such attempts as futile and to some extent is even compelled to do so in order to maintain its construction.)

Schrödinger does not renounce metaphysics in his scientific considerations.[2] He maintains it, however, only until physics is firmly established and then stores it away like a scaffold which has no permanent function. He finds that a description in physical terms with mathematical language adapted to it is not capable of dealing with our sensual perceptions. In his attempt to avoid excluding subjective feelings from the physical world he finally resorts to the philosophy of the Vedanta, as expounded in the Upanishads.[3]

The philosopher's astonishment at the existence of the world and compassion are two central motives for him. I believe that the latter is at the root of his strong endeavor to help his students and fellow man gain knowledge and a better understanding of the world. After all lack of understanding is seen as one of the greatest agonies we have to endure. This is in keeping with what he said when asked about the possible consequences of scientific progress. His concern about its potential misuse dominated over the enthusiasm for possible benefits.

The preceding serves to demonstrate how Schrödinger distinguishes himself from the prototype of today's physicists. His friend, the Spanish philosopher José Ortega y Gasset, had introduced the notion of mass man. Ironically, he considered the physicist to be a prototype of this.[4] Each one of us risks falling under this category. Schrödinger's approach to scientific thinking no doubt minimizes such a risk. His thinking is better described as comprehensive than as many-sided.

Schrödinger's thesis and his first positions required experimental work, for which he showed no particular talent. However, he believed that every physicist needs experimental experience - he even stressed to me that he considered this to be an imperative of theoretical work. His teacher Franz Exner had developed exceptional ideas in the field of physics. The heat motion to which

J. Götschl (ed.), Erwin Schrödinger's World View, 109–114.

every physical body is subject causes deviations from the purely mechanical laws of motion; the smaller its mass the more conspicuous these deviations are. Exner doubted the assumption that free microscopic particles need to follow the same laws of motion as observed for macroscopic bodies. (Classical physics does not, of course, assume that the heat motion of microscopic bodies is negligible. Rather it holds the view that during time intervals where such fluctuations become negligible, the law of motion is still that of a mechanical mass point.)

Schrödinger shared Exner's doubts, maintaining these his whole life. They also reappear in his last article.[5] The discovery of the wave properties of matter was no doubt a great triumph for Exner. Schrödinger's doubts, however, extend beyond that to the conservation laws for which no deviations have been observed.

We are confronted here with another issue that occupied Schrödinger throughout his life: The points of a continuum can hardly be attributed a physical significance. Since the beginning of science, this continuum has accompanied its new developments like an evil spell.[6]

Atomistics and quantum theory provided some relief in showing, for instance, that energy levels are discrete. Schrödinger appreciated this. He found, however, that in reality these levels have no distinct discrete eigenvalues. Thus he concluded that the presumed merit of quantum theory to emply atomistics lacks theoretical grounding. He hoped that the second quantization would help eliminate this drawback. (The inclusion of damping terms does indeed remove some unrealistic properties of energy levels and temporal development.) Schrödinger, as we know, also remained critical of the interpretations of the wave functions and the measuring process.

While I was Schrödiger's assistant, a modified view on the interpretation of quantum mechanics was developed in the United States.[7] The formalism of quantum mechanics was made the very fundamental of physics. Now measurements are the only interactions between systems dependent on the wave equation. Each measurement results in a superposition of products of wave functions of the measuring apparatus and the object of measurement. To remain consistent with experience it is assumed that this superposition of wave functions ultimately becomes (in a somewhat mystical manner) distributed over many different worlds. Only one component of the determined eigenvalue for the measuring device remains in our world. The latter component is also a partial solution of the wave equation. God does not play dice - he only contemplates all possible positions of the dice in different worlds. In our world this is how one obtains all the pertinent results of quantum mechanics without the involvement of macroscopic bodies which Bohr found necessary.[8] The branching occurs even when both systems are microscopic.[9] Observers are not needed so that even a wave function of the whole universe (likewise with somewhat mystical initial conditions) may be defined.

This interpretation was even accepted by M. Gell-Mann as he stated in a recent lecture.[10] Many features that Schrödinger criticized can be avoided through this approach. He would, however, have hardly accepted it wholeheartedly. In any case it hardly goes unnoticed that there is still the mentioned scaffold of metaphysics which reminds us that the edifice is not yet completed.

Schrödinger unfortunately took some details of his criticism with him to the grave. He tried to prevent all thesis candidates from getting too involved in these arguments. He worried in his conscientious way that we would later face professional disadvantages on the controversial subject in which his position was isolated.[11] He did not show me his last article "Might Energy be Merely a Statistical Concept"[12] until the reprints had arrived. When he asked me for my opinion I remarked that after a first reading several points in it were not yet clear to me. He retorted by citing

Faust: "Dear friend you are unaware how rude you are." He always went to great pains to achieve clarity in his writings. His knowledge of quantum statistics was, however, so extensive that the limitations of the reader may have escaped his imagination. He sometimes only alluded to considerations where detailed elaborations would have been necessary.

In an earlier article I reported details of Schrödinger's views on gravitational physics[13] which became his favorite subject during the last years of his life. I also reported there how he influenced the direction of my research. Here I am mainly interested in his views and thinking. I would like, however, to venture further at a later date even if I risk ridicule - as he encourages it - to show how the valuable remarks he made in his criticism of quantum mechanics inspired me to continue to spin Exner's yarn on my own. I may also add here that the continuum is already inherent in our subjective perceptions. The difficulties brought on by the continuum for a physical description are thus closely linked to the problem of the relation of our feelings and perceptions to physics.

In the fifties, Einstein's general theory of relativity was not yet confirmed in the solar system. Schrödinger, however, was convinced that this theory was valid in this domain. During the thirties he dealt with generalizations of the wave equations in gravitational fields. He then attempted to eliminate the continuum of energy levels by contemplating solutions of these wave equations in finite closed universes. He discovered in this connection the formation of electron-positron pairs in expanding universes.[14] After this, his great endeavour was, following Einstein, yet independent of him, to create a unified field theory to describe electrodynamics and elementary particles on the basis of general relativity. When he returned to Vienna in 1956 he already realized, through C. P. Johnson's critical observation, that his endeavors had not been successful so far.[15] He recommended me to work on the general theory of relativity and relativistic thermodynamics, but not on his papers on the unified field theory. The whole subject was at this time still foreign to most physicists. I had studied general relativity on my own out of a burning interest. At the time it seemed to be impossible to find employment in this area in Austria. My knowledge, however, was to serve me well quite unexpectedly. I received the position under Schrödinger because I was the only one whom he could discuss his favorite subject with. Unfortunately I did not follow his suggestion to work on problems of relativistic thermodynamics and statistics, as I was more interested in investigating the results of elementary particle physics, such as the violation of parity conservation and their relation to the general theory of relativity.

Schrödinger never exerted pressure on anyone to select a specific topic of research. He was always ready to discuss any question, openly voicing his opinions. He considered the desire for knowledge to be something precious - even sacred. His comprehensive knowledge and deep understanding frequently manifested itself in striking comparisons and analogies and also reflected his humour. I have compiled a number of anecdotes characteristic of him. Unfortunately, only excerpts of these have been reproduced by others without asking my consent. The reason was obviously to avoid offending anyone in influential circles. Some of his most important traits have thus been suppressed. One also has to ask whether such publications are then really honest in other aspects. I am convinced that important details are missing in the censored or extracted reproductions. (See "Remembering Schrödinger" in this volume.)

Schrödinger was convinced that the general theory of relativity describes macroscopic gravitation best. His discovery of the pair formation of electrons, which result from the expansion of the universe, made him suspicious about the application of the theory to extremely large or dense masses. The theory of relativity predicts unusual physical conditions in the vicinity of such masses. These conditions ultimately lead to the collapse of matter (and space) in a black hole. Schrödinger had already seen in classical mechanics that a well-founded theory, which was confirmed in large domains, can lose its validity under extreme conditions. General relativity, however, is constructed so as to yield the same results for suitable systems of reference in limited domains anywhere. The anomalous conditions of a black hole can, however, occur in suitably extended domains. Schrödinger did not trust the validity of the theory in such domains and repeatedly refused to discuss it with me, because the conditions were too far removed from experience.

Such a way of thinking contradicts what is generally held today. Outstanding mathematical investigations carried out by scientists such as Oppenheimer, Snyder, Israel, Hawking and Penrose have shown that for reasonable assumptions about matter distribution, the collapse into a black hole is inevitable in the general theory of relativity. A remote observer does not see the collapse within a finite time, whereas the time for an observer also moving is finite. Schrödinger already fully understood this apparent paradox. In one of his beautiful analogies he equated it to Xeno's paradox of Achilleus and the turtle.[16]

General relativity thus predicts the collapse. Astrophysical observations of pulsars and quasars indicate the existence of stars with the density of nuclear matter - up to 10^{17}g/cm^3. This is not enough density for a star of one solar mass to collapse. The majority of physicists nevertheless consider the black hole to be an established fact. (A film was even produced in which it plays a major role.)

Schrödinger already knew of the theoretical possibility of neutron stars, but he remained sceptical of the validity of any phenomena beyond our experience. He had also already considered in his unifield theory that gravity manifests itself differently in microscopic domains than predicted by general relativity. The idea of modifying the gravitational field at small distances to a meson field still appeared attractive to him even after he abandoned his unified theory. Already at an early date he was puzzled by his discovery of pair formation. If one considers that according to Riemann's fundamental insight, geometry has classical physical properties,[17] one can hardly avoid associating pair formation with time-dependent gravitational fields. Schrödinger's papers showed the effect in a most convincing manner on the basis of the solutions of wave equations in gravitational fields. Similar conditions appear in more drastic form if matter collapses locally. Geometry ultimately becomes globally non-static. After Schrödinger's work of 1939 it is hard to see how pair formation could be excluded here. Assuming, however, that it exists, it is even more difficult to exclude generalizations of the gravitational field equations. These result from virtual pairs directly related to the real pairs. The modified equations do not necessarily result in a collapse into a black hole. Investigations on the difficult subject of virtual pairs date back to 1952.[18] Even though today there is a whole industry dealing with this subject, substantial results are still lacking. In my opinion, this is due to the universality of gravitational interaction which treats all kinds of matter in the same manner. I have indicated in another paper[19] how I think this problem can be tackled.[20] Here I am only trying to show that Schrödinger had significant reasons to doubt the physical meaning of the solutions of the field equations of gravitation in extreme domains.

A decisive moment for the future direction of my research came when Schrödinger, in the last years of his life, suggested to me in a telephone conversation that he considered it more promising to attempt to gain an understanding of the magnitudes characterizing a physical theory than to study its extreme limits beyond our experience. In his opinion the most important question in gravitational theory was why the gravitational field acts so much weaker than other fields. He thought that a new idea would be required to answer this.

I consider this to be one of the wisest suggestions made on this subject. In my article[20] I point out how I oriented my research after this and how I think this question can be answered.

The beautiful mathematical results of general relativity which predict the total collapse have been considered in the meantime as physically valid, which means that the black hole is an objective reality. To demonstrate my doubts, I coined the following phrase in view of earlier considerations: "There are more things between heaven and the black hole than dreamt of in the philosophy of relativist mathematicians."

I believe that this phrase characterized the situation at the time well, because not long after, S. Hawking's famous paper on pair creation through black holes appeared. The black hole was now legalized with respect to the laws of thermodynamics while at the same time even more important virtual pairs were left unconsidered.

I have shown in my article that gravitational collapse results in the formation of additional space regions and thereby in new unoccupied quantum states for matter. The zero-point energy of these newly formed levels has to be supplied at the expense of the total energy of the system. This altered energy balance is expected to result in conspicuous modifications of the classical theory in an advanced state of the collapse. The situation is too complicated for us to be able to predict reliably what happens at such extreme stages. The clarification of this issue is an important task of observational astrophysics. It cannot today be excluded empirically that the collapse never proceeds far enough to form a horizon. We also still lack experimental knowledge on gravitational fields in the quantum domain. Only future research will be able to verify the correctness of Schrödinger's claims.

Schrödinger demanded from a physical law that it yield the same results to any observer under equal conditions. In his view of the world, he took recourse to Vedantic philosophy which appeared best suited to him to account for equal observational results of different individuals.

I would like in this context, to raise a question which leads us beyond science and academic philosophy. Meditation played a significant role at the origin of Vedantic philosophy. Roughly spoken, it can be defined as a psychological practice in which the mind should be relieved from the influence of thinking. The pictures that may appear in such a practice vary from person to person and, according to Schrödinger, do not belong to the domain of physics. He had no doubt a healthy distrust of subjective observations because natural science had unmasked so many apparent observations; he was no doubt never practicing, himself, meditation. Today meditation has spread to the western world and has become a field that psychologists and psychiatrists deal with. It triggers interesting psychological and physiological issues yet to be fully explored and explained. If, however, one adopts Vedantic philosophy as one's view of the world, it may appear adequate to also deal with such phenomena. One thereby encounters another area where metaphysics and physics meet: An experienced psychiatrist remarked that he occasionally experienced clairvoyant phenomena when meditating.[21] He advised ignoring these because following and analysing them would result in loss of the ability to meditate and, thus, also in that of finding such kinds of phenomena. I found his remarks to be essentially true. To avoid any misunderstanding, I stress that I have found neither human beings nor animals with clairvoyant capabilities which defy statistical laws or exclude explanations based on rational processes or instinct. I do not even believe that such capabilities exist.

There is also no doubt, no way to make systematic predictions about the outcome of a game of chance, and even less so of the stock market. I find, however, that clairvoyant phenomena occur sometimes as a rule spontaneously where statistical or rational explanations prove insufficient. A drastic example would be a sudden premonition of a close person dying in an unforseeable accident. The occurrence of similar events is unpredictable and there is no way to use them systematically to supercede statistical laws in many instances. The amount of new information in each case is, however, so great that statistical fluctuations can practically be excluded to account for it. There remains of course, the question whether they really are genuine. The answer to it may remain for a long time a matter of personal experience. One should also not ignore the fact that there is a strong drive to produce such events.

There are also important domains of human life for which Schrödinger did not show much understanding. Politics no doubt was one of them. When he left Berlin in 1933 he had no doubt recognized that Hitler was becoming the greatest danger to German culture. He did not, however, forsee the next developments.

He finally accepted, after renouncing a position at Oxford, a professorship at the University of Graz. At the end of February 1938, when Austria's fate was already sealed, he held a public lecture in Vienna in which he stated that when contemplating the universe, the claim of one nation or race to be superior to others must appear ridiculous. When Hitler integrated his country of origin in his insane ventures, Schrödinger, of course, fell in disgrace and had to experience duress. He told me that he made this declaration overly enthusiastically so that his friends abroad could recognize that it was not genuine. After an Academy meeting in Italy he did not return to his home country. It was not until 1956 that he moved back to Austria.

NOTES

1 E. Schrödinger, *Science and Humanism*, Cambridge 1952.
2 E. Schrödinger, *Meine Weltansicht*, Paul Zsolnay, Vienna 1961.
3 E. Schrödinger, *Science and Humanism*, op. cit.
4 ibid.
5 E. Schrödinger, *Il Nuovo Cimento* (10) 9, p. 162 (1958).
6 E. Schrödinger, *Science and Humanism*, Cambridge 1952.
7 H. Everett, *Rev. Mod. Phys. 29*, p. 454 (1957).
8 J. A. Wheeler, *Rev. Mod. Phys.* 29, p. 463 (1957).
9 ibid.
10 M. Gell-Mann, lecture held at the 5th Marcel Grossmann Meeting, August 8, 1988, Perth, Australia.
11 H. Pietschmann, in: *Dokumente, Materialien und Bilder zum 100.Geburtstag von E. Schrödinger*, p. 135, G. Kerber, A. Dick and W. Kerber (edit.Fassbaender) Vienna 1987.
12 E. Schrödinger, *Il Nuovo Cimento* (10) 9, p. 162 (1958).
13 L. Halpern, *Found. of Physics 17*, p. 1113 (1987).
14 E. Schrödinger, *Physica 6*, p. 899 (1939) and Proc. Roy. Irish Acad. A 46, p. 25 (1940).
15 C. P. Johnson, *Lett. Phys. Rev. 89*, p. 320 (1953).
16 E. Schrödinger. "Expanding Universes", Cambridge (1957).
17 L .Halpern, *Ann. of Physics 25*, p. 387, 1963.
18 B. De Witt, thesis at Harvard University, 1952.
R. P. Feynman, *Lectures on Gravitation*, Caltech 1971.
S. N. Gvpta, *Proc. Phys. Soc. A 65*, p. 161 and p. 608, 1952.
19 L. Halpern, *Found. of Physics 17*, p. 1113, 1987.
20 ibid.
21 L. Le Shan, *How to Meditate*, Bantam, New York 1974.

ERWIN SCHRÖDINGER AS HISTORIAN - NOTES TOWARDS AN INTERPRETATION

YEHUDA ELKANA
The Institute for the History and Philosophy of Science, Tel Aviv University and The Van Leer Jerusalem Institute Albert Einstein Square, 91040 Jerusalem, Israel

1. INTRODUCTION

Erwin Schrödinger was undoubtedly one of the greatest physicists of our time, if not of all times. He was a philosopher who dealt with metaphysical and existential problems and tried to connect his philosophical views with his physics. And even if this is disputed by some, he was a central force behind the emergence of modern molecular biology. On the other hand, Schrödinger is almost never referred to as a historian, even though he explored the history of Western culture and wrote a remarkable book on the Greeks.

If a historian is characterized by the ability to view all events, processes, and even positive contributions to knowledge from a relativistic, context-dependent, comparative standpoint, then Schrödinger was a historian par excellence. He was humanistic in the classical, sophistic sense; in comparing cultures, he saw man and his intellectual products as being determined by context. His cognitive psychology was culturally conditioned, and his fundamental approach evolutionary. There is something seemingly paradox to this: as a physicist Schrödinger was one of the creators of a new physics; his humanist outlook and historical approach to the world were very much shaped by his background, cultural milieu and generation. Yet the paradox is only "seeming". Schrödinger's physics also reflected this humanist, classical approach. In a different sense than Bohr's, Schrödinger's physics, too, constitutes a direct continuation of classical physics. A longer study could show in more detail the intellectual and cultural links between Schrödinger and Wittgenstein, but also ones that are less obvious, i.e., the affinities to the Schönberg School of music, the Vienna expressionists, and the "uprooted" geniuses of language like Franz Kafka and Karl Kraus or the almost dissonant relation with the work of Josef Weinheber.
(Since it would go beyond the scope of this paper to analyze this, let me mention only a recent article by J. P. Stern: "Worte sind auch Taten." Bemerkungen zum Österreichischen Sprachbewußtsein." Merkur 43 (1989), p. 673 - 687.)

2. PHYSICS: THE SOURCE OF EVERYTHING?

Schrödinger's complex, relativistic, context-oriented, science-based and at the same time broadly humanistic outlook, is best reflected in the four public lectures ("Science as a Constituent of Humanism") he held under the auspices of the Dublin Institute for Advanced Studies at University College, Dublin, in February 1950. These lectures were published by Cambridge University Press in 1952 as *Science and Humanism: Physics in our Time.*

J. Götschl (ed.), Erwin Schrödinger's World View, 115–134.

What is so striking about this little book are its far-reaching philosophical implications that are in keeping with what he writes in the preface: "I see scientific work as forming part of man's endeavour to grasp his situation."[1]

Schrödinger rejects any artificial division of culture and tries to find answers to the Plotinian, "And we, who are we anyhow?"[2] He looks for an answer in science, in epistemology, in a context: "I am born into an environment - I do not know where I came nor whither I go nor who I am."[3] He also tries to describe the way science should be taught so as to enable its students to learn from it in order to be able to tackle existential problems. He deplores specialization as a necessary evil, and admonishes all scientists to "never lose sight of the role your particular subject has within the great performance of the tragicomedy of human life."[4]

But what Schrödinger wants to extract from science are deep insights into the nature of the world, not moral lessons or conclusions about free will. One can easily see a reductionist scientific approach in this. Surveying what modern physics teaches us about the possibility of continuous observation (dismissed by Schrödinger), causality, and the relationship between subject and object (he still sees a boundary between subject and object), Schrödinger draws an anti-reductionist humanistic conclusion:

> For the observing mind is not a physical system, it cannot interact with any physical system.[5]

Already here the deep tension between contradictory approaches, prevalent throughout all of Schrödinger's work, is apparent.

Schrödinger then deals with the vexing problem of free will. Referring to the old paradox, or *aporia*, of how we can be responsible for our laws, or God-predeterimined acts, he raises the question whether *free* can come to hear the way that *free will determines* those events which the Law of Nature leaves undetermined?[6] This was Pascual Jordan's solution and Schrödinger finds it "physically and morally an impossible solution."[7] The moral impossibility is more interesting: following Cassirer, Schrödinger admits that "the mental phenomenon of will, ... is ... usually *extremely* determined from inside."[8] So free will, including ethics, is strongly determined and cannot be based on physical coincidence. Schrödinger concludes his book by quoting Cassirer:

> Thus it is clear ... that a possible change in the physical concept of causality can have no immediate bearing on this.[9]

In 1936 Schrödinger had published in "Nature", a paper on "Indeterminism and Free Will."[10] Here, in a very concise way, he takes up the same views again as those addressed in the 1952 book discussed above.

To the first 1944 edition of *What is Life*?, which was based on the 1943 Dublin lectures, Schrödinger added an epilogue: "On Determinism and Free Will." Here he explores the scope of the 'I', the conscious mind, which is never experienced in plural with regard to the world: "Within a cultural milieu (Kulturkreis) where certain conceptions have been limited and specialized, it is daring to give to this conclusion the simple wording that it requires."

The other work by Schrödinger which is important for understanding his view of the world, is his *My View of the World*.[11] We shall discuss this book below.

Thus both science and human destiny are products of man. Therefore there is no great divide between nature and man. Both are constructs of the human mind. Each, however, should be seen as determining the other.

3. SCIENCE AS DEPENDENT ON CULTURAL MILIEU

Schrödinger's most profound analysis of science as a product of culture, and its context-dependence can be found in his 1932 lecture, "Ist die Naturwissenschaft milieubedingt?"[12] Following Zola's "L'art c'est la nature vue au travers d'un tempérament"[13], as far as the humanities are concerned, Schrödinger raises the question whether indeed, as is often claimed, the opposite is true for science. Does the Sophistic "man the measure" break down in science? According to Schrödinger the distinction "goes too far". Considering that in spite of the fact that science is experimental, we do not perform all possible experiments, makes the result very human indeed:

> Therefore a selection has been made in choosing the raw material on which the present structure of science is built. The selection must have been influenced by circumstances that are other than purely scientific.[14]

Issues he focuses on in his non-scientific considerations include, practicality on the one hand, and personal interest, presuppositions, etc. on the other. Some very illuminating historical examples follow, showing how Schrödinger constructs a historical argument. All in all science is, like sports, "milieu-determined". Sports are simply a "Modesache" (fashion). In the German original:

> Ob man sich aber fuer Hochsprung, Tennis oder Diskus oder fuer andere in Zukunft frei zu erfindende Spiele interessiert, ist Modesache. Wie sollen wir aber den Beweis fuer die starke Auspraegung subjektiver Momente in der Zielsetzung der Wissenschaft erbringen? Das ist nicht ganz leicht. Man steckt doch schliesslich in seiner Zeit drinnen. Wenn Modeanschauungen vorliegen, so teilt man sie.[15]

And his summary:

> Unsere Kultur bildet ein Ganzes. Auch wer das Glueck hatte, die Forschung zu seinem Hauptberuf zu machen...liest Romane und Gedichte, geht ins Theater, treibt Musik, macht Reisen, sieht Bilder, Skulpturen, Architektur - und vor allem, man liest und spricht viel ueber diese und andere Dinge. Kurz wir alle sind Mitglieder unseres Kulturmilieus.[16]

Schrödinger explores the symbolic representations of knowledge which constitute our cultural milieu and influence our science. He knows that he is not describing the world but rather our thought-processes (i.e., not the body of knowledge but rather images of knowledge), as seen in the following quotation:

> ...mit der Kausalitaet genau so steht wie mit der Geometrie. Experimentell wird sich niemals entscheiden lassen, ob Kausalitaet in der Natur "gilt" oder "nicht gilt". Das Verhaeltnis der Abhaengigkeit von Ursache und Wirkung ist, wie schon Hume erkannt hat, nicht etwas, das wir in der Natur vorfinden, sondern es betrifft die Form unseres Nachdenkens ueber die Natur.[17]

4. THE QUEST FOR UNITY

a) The Universal Context

In Born's Festschrift, written in 1940 but not published until 1953, Schrödinger makes it clear that his greatest objective is unity: all forces are in a sense gravitational and one must add wave mechanics as a complement to relativity theory.[18] This is the basic context of Schrödinger's thinking. It underlies his views on causality and the statistical nature of the world, as well as his insights into biology. It is also the link with his own cultural milieu: a torn world, suffering from an almost unbelievable tension between facade and truth, believing only in an indeterministic epic historical process, held together by a fundamental unity of the world of man.[19] Schrödinger's dealings with Boltzmann and Einstein and his views of their theories can only be understood within this more complex context.

Schrödinger studied in Vienna starting in 1906 shortly after Boltzmann's death. His teachers Hasenöhrl and Exner were both Boltzmann's students. Schrödinger followed Boltzmann's realism, statistical thinking, admiration for Darwin, and his epistemology.[20]

In 1891 Franz Exner was appointed to succeed Loschmidt in the chair for experimental physics, and in 1908 became rector of the University of Vienna. In his inaugural lecture,[21] in all probability attended by Schrödinger, then a second year student, he first gave his famous explanation that all laws of physics, even classical physics, are statistical averages. In his 1919 textbook he dedicated several chapters to this issue, and Schrödinger referred to the idea quoting Exner in his own inaugural lecture in Zurich in 1922. For Exner, there are no "two cultures": all phenomena, even those of history or economy are natural phenomena and, therefore, subject to statistical laws of probability. Only the huge number of events among the phenomena of non-human nature make their laws more exact than in humanistic fields. According to Exner it was Boltzmann who showed:

> ...that the world constantly moves from a less probable to a more probable state and thus to an increasingly stable state.

Exner refers to an idea which is recurrent in Schrödinger's work: chaos among mere chance events before the statistical probability of large numbers.

Paul Hanle studied Exner's ideas and their influence on Schrödinger:[22] he emphasizes the similarity in philosophical outlook between Exner and Schrödinger to the extent that both "publicly supported social and philosophical beliefs with physical principles including fundamental indeterminancy, while in turn they granted that their beliefs had conditioned these principles".[23]

In 1944 Shrödinger returns to Boltzmann explaining that it was due to Boltzmann and to Gibbs that statistical thinking entered science, not just as a "remedy" to our lack of knowledge of details but as the conceptual basis of all laws of nature:

> ...they are not rigorous laws at all, but 'only' statistical regularities based on the law of great numbers, just as Darwin's theory is.[24]

Not only does statistics *not* serve as a remedy for our ignorance of details about individuals, it also replaces our need for such detail. We learn through statistical thinking new facts about the world which one would not even know if one knew all the details:

> It dawns upon us that the individual case is entirely devoid of interest whether detailed information about it is obtainable or not, whether the mathematical problem it sets can be coped with or not.[25]

Boltzmann, in 1886, suggested that the 19th century will be called the century of Darwin, and Schrödinger agreed with this view. This view is historical in the same sense that Darwinian evolution is history.

Dieter Flamm[26] brought together the various areas of physics where Boltzmann's influence on Schrödinger was directly or indirectly detectable. But the influence of culture on statistical thinking is much more complex than the local Viennese network of Boltzmann, Exner and

Schrödinger. Exner experimented with the Brownian notion in 1900, five years before Einstein's epoch-making paper. An in-depth study showing the deep historical links between fin-de-siècle culture and statistical thinking is overdue. Boltzmann's influence was also mediated via the series of lectures in physics by Fritz Hasenöhrl ("eight semesters five hours a week!")[27] and Franz Exner with whom Schrödinger worked .

Exner's 1909 Rektoratsrede renders what the debates focussed on at the time:

...we were able to slow down the molecular motions so enormously that we could follow the individual molecular processes we would search vainly for lawfulness.[28]

Schrödinger's 1950 paper, "Irreversibility," is dedicated to one of Boltzmann's most critical problems: can we arrive at a satisfactory explanation of irreversible behavior, which is *de facto* most of nature, from reversible models, which constitutes the greatest part, if not all of physics? Schrödinger states:

I wish to reformulate the laws of phenomenological irreversibility...in such a way that the logical contradiction which any derivation of these laws from reversible models seem to involve is removed once and for ever.[29]

Thus physics and history move closer together: history is always irreversible but it does not try to explain events with a reversible model - not even a cyclic theory of history is strictly reversible. Moreover, Exner:

...nicht jedes Geschehen moeglich ist, und die Erscheinungen etwa ebenso gut wie anders haetten verlaufen koennen.[30]

This indeed is the *epic view* [31] of history and not the determinist's dramatic view! This indeterministic approach and the strong influence of cultural unity and of course of the physical world also shaped his scientific research program. It is quite in keeping with what has been said that Schrödinger became one of the first cosmologists to look for the key to the cosmos in elementary practical theory.

Schrödinger's relations with Einstein are a bit puzzling.[32] On the one hand, Schrödinger clearly aligned himself with Einstein against those who rejected a clear physical representation of the world and called for only mathematical precision in the material world. Both of these two great thinkers who shaped our view of the world, never accepted the Copenhagen interpretation in its Bohr-Heisenberg version. On the other hand, while Einstein viewed the world as fundamentally deterministic, Schrödinger's view was that both classical and quantum worlds are basically statistical.[33]

Schrödinger's contextualism also forms the basis of his original view on causality and determinism. As already noted, Schrödinger, like Einstein, opposed the prevalent readiness to give up an objective description of physical reality and was thus very much a physical realist. But unlike Einstein, he did not accept a strict deterministic causal approach as a natural consequence of realism. He is nowhere more explicit about this than in the abstract written for a strictly scientific paper which was published in 1958 by the Austrian Academy of Sciences:

...the widespread attidude that the claim for an objective description of physical reality must be given up is rejected on the grounds that the so-called external world is built up exclusively of elements of the single minds, and is characterized as what is common to all, recognized by every healthy and sane person. Hence the demand for a non-subjective description is inevitable, of course without prejudice whether it be deterministic or otherwise.[34]

Schrödinger's statistical world view automatically leads us to the question whether he accepted a fundamental causal view of the world or not. History, too, can be viewed causally or acausally according to the historian's metaphysics, but those historians who want to see history as a causally determined chain of events generally wish to see history reduced to laws similar to those of physical nature. On the other hand, historians who think in acausal terms very often tend to view history as quite different from physical science which, according to most, is strictly causal.

Following Exner, Schrödinger rejected the view that the laws at the atomic/molecular level are determinate-causal. Laws governing the impact of molecules are indeterminate and ruled by "an appropriate law of chance." [35]

Exner extrapolated from microphysical phenomena to the macro level - for him all laws of nature are inexact and statistical in character. Exner actually called laws of nature "habits of thought." [36]

In his 1922 inaugural lecture in Zurich, Schrödinger describes causal thinking as a *custom*.[37] Schrödinger summarizes Exner's views and his agreement with them as follows:

Exner's assertion amounts to this: it is quite possible that Nature's laws are of thoroughly statistical character...*The burden of proof falls on those who champion absolute causality, and not on those who question it.*[38]

Significantly, Schrödinger finds a formulation of Relativity Theory that does not clash with Einstein's causal thinking, yet fits his own acausal philosophy:

For Einstein's gravitation theory is not really anything more than the reduction of gravitation to the law of inertia. That under certain conditions *nothing changes* is surely the simplest Law that can be conceived, and hardly falls within the concepts of causal determination. It may after all be reconcilable with a strictly acausal view of Nature.[39]

Schrödinger, like Exner, was rooted in his own cultural milieu which conceived of life and the world in epic terms: it is not predetermined that what happened had to happen. Thus the historical question is not the dramatic "how did it come about that what had to happen anyhow," but rather "since anything could have happened why did this happen?" [40] Thus even:

Das tiefe Problem der Kausalitaet scheint mir doch in Folgendem zu liegen: Warum erwarten wir unter *voellig* gleichen Umstaenden stets auch den *voellig* gleichen Erfolg?...[41]

It is no surprise that Schrödinger welcomes with enthusiasm the 1924 Bohr-Kramers-Slater paper suggesting an acausal microphysics by considering even the conservation of energy a statistical average which does not apply strictly to individual particles.[42] When, in view of most physicists, the Bothe-Geiger and Compton-Simon experiments convincingly refuted the BKS theory, Schrödinger continued to stick to it by interpreting the Compton effect in a different way.[43] And he returned to his 1927 interpretation and argued the same case from a different point of view, once again in 1958.[44]

Schrödinger, like Exner, was rooted in his own cultural milieu which conceived of life and the world in epic terms: it is not predetermined that what happened had to happen. This historical

question is not the dramatic "how did it come about that which had to happen anyhow," but rather "since anything could have happened why did this happen?"[45]
Thus, even:

Das tiefe Problem der Kausalitaet scheint mir doch in Folgendem zu liegen: Warum erwarten wir unter *voellig* gleichen Umstaenden stets auch den *voellig* gleichen ^Erfolg?...[46]

For Schrödinger, causality is a symbolic representation of knowledge, a frame of mind:

Es handelt sich bei dieser Frage meines Erachtens nicht um eine Entscheidung ueber die wirkliche Beschaffenheit der Natur, wie sie uns entgegentritt, sondern ueber die Zweckmaessigkeit und Bequemlichkeit der einen oder der anderen Einstellung unseres Deutens womit wir der Natur gegenuebertreten.[47]

A few lines later we read:

Es sind wohl kaum Erfahrungstatsachen denkbar, welche endgueltig darueber entscheiden, ob das Naturgeschehen in Wirklichkeit absolut determiniert oder partiell unbestimmt ist, sondern hoechstens darueber, ob die eine oder andere Auffassung einen einfacheren Ueberblick ueber das Beobachtete erlaubt.[48]

In the above-mentioned 1932 lecture on the cultural context of science, Schrödinger repeated the claim that causality is a "symbolic representation of knowledge" with even greater clarity.

Das Verhaeltnis der Abhaengigkeit von Ursache und Wirkung ist, wie schon Hume erkannt hat, nicht etwas, das wir in der Natur vorfinden, sondern es betrifft die Form unseres Nachdenkens ueber die Natur.[49]

Einstein, too, was influenced by Hume, but he physics, i.e. the world, not our knowledge of it.

b) The Biological Context
For Schrödinger, the "keen longing for unified, all-embracing knowledge"[50] is an evolutionary phenomenon. We inherited from our forefathers a desire for *universals*. The belief is reductionistic in the sense that our inability to account for the big questions in terms of physics is no reason to doubt that they can be so accounted for! Schrödinger sees a clue to future developments in viewing the living cell as an *aperiodic crystal*.

Schrödinger spoke out often, and with conviction, on the science of life, and his small book *What is Life?* (based on lectures delivered under the auspices of the Dublin Institute for Advanced Studies at Trinity College Dublin, in February 1943) had an enormous impact and was widely discussed. There is no historian of modern biology who does not mention Schrödinger together with Delbrück as the physicist who influenced the beginning of molecular biology. Yet many historians and scientists go out of their way, rather angrily, to state that Schrödinger had nothing new or even anything at all to say about biology. Linus Pauling, for example, states:

It is justified, in my opinion, to say that Schrödinger, by formulating his wave equation, is basically responsible for modern biology.

We might now ask the following question: To what extent...did Schrödinger contribute...to our understanding of the nature of life? It is my opinion that he did not make any contribution whatsoever, or that perhaps, by his discussion of 'negative entropy in relation to life' he made a negative contribution.[51]

Max Perutz is even more critical with regard to *What is Life?*:

...what was true in his book was not original, and most of what was original was known not to be true even when the book was written.[52]

It is certainly not up to the historian of ideas to take a stand in such a debate. Rather, let us try and see what is peculiar, original and important in Schrödinger's approach to biology. It was the originality of his approach which made his impact so great, irrespective of the truth value or the specific claims.

Schrödinger, committed to the unity of knowledge, aimed at:

...to embark on a synthesis of facts and theories, albeit at the risk of second-hand and incomplete knowledge of some of them - and at the risk of making fools of ourselves. So much for my apology.[53]

Physics as it was in 1943 had no tools to deal with biological phenomena. The laws of physics and chemistry were statistical throughout and:

...it is in relation to the statistical point of view that the structure of the vital parts of living organisms differs entirely from that of any piece of matter that we physicists and chemists have ever handled physically in our laboratories or mentally at our writing desks.[54]

The most complicated object that physics can deal with is a *periodic crystal*. On the other hand:

The essential part of a living cell - the chromosome fiber - may suitably be called an *aperiodic crystal*.[55]

But this is the point. Schrödinger was not only very familiar with Delbrück's work, but it is well-established that only Schrödinger's chapter 5: "Delbrück's Model Discussed and Tested" in his *What is Life*? drew the attention of the scholarly community to Delbrück's idea which indeed contributed directly to biology according to both Pauling and Perutz. The fact that Schrödinger calls it not a polymer but an aperiodic crystal shows where his interest lies: not in chemistry proper, but in the epistemology of research. The range of scientific thinking had to be widened that it could cope with aperiodic phenomena; only then could knowledge be united. Aperiodicity has to do with history. Schrödinger's example is illustrative in this context:

The difference in structure is of the same kind as that between an ordinary wall-paper in which the same pattern is repeated again and again in regular periodicity, and a masterpiece of embroidery, say a Raphael tapestry, which shows no dull repetition but an elaborate, coherent, meaningful design traced by the great master.[56]

In other words, biology, in order to be studied 'scientifically' would have to 'interpret' in the same way as history does: discovering meaningful design in aperiodic phenomena. Here again we confront the symbolic representation of knowledge typical of Schrödinger: unity of knowledge, and thus no great divide between the humanities and natural sciences. That unity was to be attained through one common approach - what he then calls science will take over all phenomena: natural and human, periodic and aperiodic. This may sound positivistic-scientistic. However, the kind of science that was to be developed in order to cope with all phenomena was a historically relativistic, context-oriented, hermeneutic type of science. And statistical thinking was supposed to help us find something uniting all phenomena and to discover the invariants among aperiodic phenomena. The universe as a whole was what he has in mind.

Schrödinger's thermodynamics and his vague definition of what constitutes a system has been criticized by Pauling. But this is not what preoccupies Schrödinger. He is interested in showing biological phenomena that are specifically relevant for the new physics. It is the fact that organisms feed on order which they absorb from their environment, so as to counteract their own disorder-creating processes which all lead to death. They are alive as long as their absorption of order is greater than their creation of disorder:

> Thus the device by which an organism maintains itself stationary at a fairly high level of orderliness (=fairly low level of entropy) really consists in continually sucking orderliness from its environment.[57]

If there is a strong residue of 'ignorabismus' which is almost mystically hidden from us, it is in our inability to apply our own ideal new science to our own thinking. This idea is expressed twice in quite different formulation:

> If it were not so, if we were organisms so sensitive that a single atom, or even a few atoms, could make a perceptible impression on our senses - Heavens, what would life be like! To stress one point: an organism of that kind would most certainly not be capable of developing the kind of orderly thought which, after passing through a long sequence of earlier stages, ultimately results in forming, among many other ideas, the idea of an atom.[58]

In sum, two claims have been made: first, Schrödinger's aim to unite the whole universe, or to subsume our whole world of knowledge of the universe to new laws which apply to all phenomena, is statistical in character, supplying a 'context-dependent' hermeneutic for the aperiodic phenomena.

Second, that Schrödinger's influence went far beyond having discovered wave-mechanics which then influenced chemistry and, through it, biology. The claim made by Schrödinger's critics, admiring as they may be, namely that his influence on the life sciences would have been the same or possibly even greater had he not addressed biology as wrong. It is the programmatic character of *What is Life?* and many of the epistemological asides have proven to be wrong. Some of them should have been recognized as being wrong already in 1943. By drawing attention to Delbrück's work, this little book encouraged the study of non-isotropic and non-periodic phenomena, and emphasized the claim that biology must and will be reduced to physics, but to a different type of physics than known at his time.

5. THE INDIVIDUAL IN A PHILOSOPHICAL CONTEXT

We started out by looking at Schrödinger's claim that science, or rather that knowledge, is dependent on cultural milieu. We touched upon his preoccupation with and attitudes to the problem of free will. Then we considered the main body of Schrödinger's scientific oeuvre and tried to show how it was dependent on a specific milieu, i.e., the context in which Schrödinger matured and acted. Now, bringing the cycle full circle, we shall look at Schrödinger's approach to the individual, his language, psychology, and how this again is greatly influenced by cultural milieu and morals.

In 1954 Schrödinger participated in one of the famous Jungian Eranos meetings. His topic was "The Spirit of Science" and he returned to several of his favorite topics, dealing with these in various ways. Here, again, it becomes clear that Schrödinger is a historically-minded, context-oriented philosopher of nature. But his historical approach, as can be seen in his emphasis of context, stands in opposition with, and is sometimes dulled by, a quasi-positivistic, quasi-scientistic quest for clarity and sharp distinctions which are inherently ahistorical and non-relativistic:

The spirit is to an eminent degree subject, and thus evades objective examination. It is the subject of cognition (Schopenhauer) and therefore, strictly speaking can never be its object.[59]

As usual, whenever Schrödinger mentions Schopenhauer he goes on to link the idea with Vedanta philosophy. Following the English indologist (of German origin) Max Müller, he considers the '*We*' or 'I' in the Vedanta-sutras as the subject, which cannot be perceived as such, whereas the 'Non-I', the '*You*', the objective, everything what is known of science belongs to it. Great philosophers of India were simply not interested in the objective universe, they were interested in the knowing, thinking subject, which has exerted a great fascination on thinkers in the western world. This interest is considered a historical fact, but not seen as an universal imperative. However, the distinction itself is not considered to be context-dependent, historically conditioned happenstance, but as logically derived, clearly conceived two alternatives. In Schrödinger's thinking there is no place for "metis", the cunning reason, the sophistic, and later Baconian conception that the kind of answer you get from "nature" depends on how you formulate your question. On the other hand, it is also unquestioned by Schrödinger whether we can get any knowledge about the knowing subject.

This very sharp distinction is, in my opinion, inappropriate, both for the physical sciences and for the cognitive sciences and constitutes a major dialectical tension in Schrödinger's thinking. Especially the question of consciousness is one that binds subject and object in a way that the above dichotomy cannot hold. If, as Schrödinger says;

The object of all science is nature in the broadest sense, i.e., our spatial and temporal environment in all its aspects. The subject of every science is always the spirit...But the sciences are a product of the spirit in which they are conducted.[60]

The question is then where our environment - our non-ego and our ego - begins and where it ends.

Schrödinger claims:

Die Welt ist ein Konstrukt aus unseren Empfindungen, Wahrnehmungen, Erinnerungen.[61]

Yet he rejects the possibility that without consciousness the world would be:

...ein Spiel vor leeren Baenken geblieben, fuer niemanden vorhanden, und darum recht eigentlich *nicht* verstanden...[62]

These are very strong constructivist statements. At the same time, in a 1931 interview, he says:

Consciousness cannot be accounted for in physical terms. For consciousness is absolutely fundamental. It cannot be accounted for in terms of anything else.[63]

Cognitive psychology and anthropology make the point that consciousness is not an individual, subjective phenomenon but rather a social one. This is the main thesis of the whole corpus of Wygotskyan psychology, much of the work of Jack Goody, David Premack, Esther Goody, Michael Cole and many others. This approach could have mediated between his constructivism and "scientistic" approach to consciousness, but it was not viable for Schrödinger. His strong individualism and his emphasis on cultural milieu prevented him from having socio-anthropological insights. Aware of this, he looked for new approaches elsewhere.

For Schrödinger, as for Exner, history and physics were both on a continuum with different degrees of regularity. Just as physical laws are statistical and never exact, history, too, is not altogether haphazard. Or at least, when so presented it is not interesting. In his short autobiography, he states:

Ich war ein guter Schueler, ohne Unterschied der Faecher, liebte Mathematik und Physik, aber auch die strenge Logik der alten Grammatiken, hasste nur das Memorieren der "zufaelligen" historischen und biographischen Daten und Tatsachen.[64]

And again:

For the observing mind is not a physical system, it cannot interact with any physical system. And it might be better to reserve the term "subject" for the observing mind.[65]

Schrödinger's original views of *continuity* resulted from this discussion on object/subject. The most lucid discussion of continuity is to be found in his 1950 lecture on "Science and Humanism": the classical ideal of the *postulate of continuity of description* (or of observation) according to which we would know precisely what happens at any point in space at any moment of time no longer holds. There are, as it were, gaps in our picture of reality. This meant the end of classical causality. The epistemological 'way out' of this impasse turns out to be very complex:

So the facts of observation are irreconcilable with a continuous description in space and time. What is to be done? The method adopted at present may seem amazing to you. It amounts to this: we give a complete description, continuous in space and time without leaving any gaps, conforming to the classical ideal - a description of *something*. But we do not claim that this 'something' is the observed or observable facts...

But what is the use of such a description, which, as I said, is not believed to describe observable facts or what nature really is like? Well, it is believed to give us *information* about observed facts and their mutual dependence. It is an optimistic view, viz. that it gives us *all* information. It is pessimistic in another respect, we might say epistemologically pessimistic. *For the information we get as regards the causal dependence of observable facts is incomplete.* This ambiguity is essential, it cannot be helped. This is the logical situation at present.[66]

Elsewhere, Schrödinger connects them with Plotinus: "And we, who are we?"

It seems plain and self-evident, yet it needs to be said: the isolated knowledge obtained by a group of specialists in a narrow field has in itself no value whatsoever, but only in its synthesis with the rest of knowledge and only in as much as it really contributes to this synthesis something toward answering the demand (who are we?).[67]

Always in a reflective vein, Schrödinger draws our attention to the fact that major changes (Umbildung) come from two contrary directions: the atomic picture was a response to the criticism that the previous approach was much too continuous, while wave-mechanics developed following, (not due to), the criticism that the classical picture was much too discontinuous.

The seeming discrepancy between continuity and discontinuity is removed by Schrödinger through an epistemological move: our ontology gives a discrete picture while our epistemology is continuous. In other words, the objects of our knowledge are discrete, while the continua of our cognition supply the relation between subject and object. And yet, is it not the case, asks Schrödinger, that our knowledge, i.e., the relation between subject and object, is the only reality?

> Warum muessen wir durchaus uns selbst ausschalten, hat nicht Gott selber uns durch den Mund unseres Dichters sagen lassen, dass wir es sind, die Ordnung in seine von Haus aus etwas chaotische "Natur" bringen sollen...[68]

As startling as this conclusion sounds, it is a solution to the quandary. Reality is second-order, reality is to be found in the symbolic representation of knowledge, i.e., continuity.

That the metaphysical debate on world views tends to confuse those two different levels is not surprising, nor is it unusual; the same thing actually happens in the Einstein-Bohr debate: for Einstein the world can behave in any imaginable way as long as our knowledge of it is deterministic; for Bohr our knowledge can be deterministic or not, as long as the world acts like proper approximation, classically. In other words, Einstein's ontology is open to anything that may turn up in science as long as his epistemology can be deterministic. Bohr's epistemology is open to any development as long as his ontology is classical. Of all the great physicists who have contributed to a new view of the world, Schrödinger is unique because he is fully aware of the distinction between the first-order and the second-order thinking, between the body of our knowledge and representations of knowledge.

6. THE RELATIVISTIC WELTANSICHT: SCHRÖDINGER'S REFLECTIVITY

Schrödinger's historical relativism as a metatheory for science, in the sense of Schopenhauer, leads us again to the 1932 essay: "Ist die Naturwissenschaft milieubedingt?"[69] There Schrödinger reiterates the producibility condition for all experiments as objective facts. Yet he rejects the classical positivist ideal that not only the individual person as object must be eliminated from science but even man as species must disappear as a subject of research. Subjective, culture-dependent factors enter the picture in the selection of topics for experiments according to community interests, but more in psychological, rather than in anthropological terms. The issues selected are determined by culture. Regarding the Kirchhoffian claim that only exact description has any scientific value, Schrödinger says:

> Erkenntnispsychologisch ist es aber einfach nicht wahr. Es ist nicht wahr, denn die quantitativen Gesetzmaessigkeiten, die wir feststellen, uns *an sich* irgend im geringsten interessieren...Sie interessieren uns ganz ausschliesslich um der Gedanken willen, die wir uns darueber zu machen beabsichtigen...[70]

The characteristics of modern physics are culture-conditioned, or rather shared by the whole of culture. This context is described as follows:

1) the "pure objectivity" as formulated in the visual arts and
2) the strong need or desire for revolution or intellectual upheaval accompanied by a love for freedom and lawlessness

3) the idea of relativity with the concomitant theory of invariants
4) a method of mastering mass-phenomena ("Methodik der Massenbeherrschung, teils durch rationelle Organisation, teils durch fabrikmaessige Vervielfaeltigung")
5) statistics[71]

This "radical approach" is typically Viennese; it is more an expression of our ideas (thus a symbolic representation of knowledge) than an ontological statement. Relativity is seen as relativism, namely a point of view which can see all truth-claims as context-dependent. Invariants are the answer to our question as to what remains universally non-relativistic.

The idea that "New Objectivity" is typical of modern physics is probably the most original and far-reaching one. The way Schrödinger understands it is as follows: unadorned large spaces, like walls or bare furniture tops, can be seen as analogous to realms of our knowledge where we have nothing to say in a strong observation-rooted sense; instead of cluttering up such areas with senseless speculation we should leave them as plain surfaces - as gaps in our knowledge!

As we have seen above, Schrödinger moves freely between ontology and epistemology. He moves from the body of knowledge to images of knowledge without even pausing. We saw that in another context. Now we are led back to physics. Having expressed his dissatisfaction with Dirac's warning ("beware of forming models or pictures at all!"),[72] he now discusses the Heisenberg principle in a very idiosyncratic way:

> If Heisenberg's assertion is to be correct, and if it appears at first sight to make gaps in our picture of the world which cannot be filled, then the obvious thing to do is to eliminate the regions which refuse to be filled with thought;...the regions in question are not certain domains in space and time (to put it bluntly, the interiors of the atoms!) but domains of abstract thought...the elimination ought to be possible without leading to the consequence, that no visualizable scheme of the physical universe whatever will prove feasible.[73]

On the one hand, he exhibited the "Neue Sachlichkeit" in physics by admitting the existence of empty regions - gaps - in our abstract thought. Yet, on the other, he stuck to the very realistic criterion of visualizability in physics.

Schrödinger's only historical work is the *History of Greek philosophy.*[74] We shall not examine it here in detail, suffice it only to say that the approach to Greek philosophy itself is not really historical. On the other hand, Greek philosophy is viewed as representing the historical origins of modern science. Schrödinger wanted to show that:

> fundamental features of the present-day scientific world-picture...are historically produced (as against logically necessitated)...[75]

A late philosophical paper on the foundations of physics opens with a section titled: "The Cultural Background."[76] With deep historical insight Schrödinger views the

> ...innovations of thought in the last 50 years, great and momentous and inavoidable as they were,...usually overrated compared with those of the preceding century;...disregard for historical linkage...is a tendency to forget that all science is bound up with human culture in general, and that scientific findings, even those which at the moment appear the most advanced and esoteric and difficult to grasp, are meaningless outside their cultural context.[77]

And again:

> Science is not a soliloquy. It gains value only within its cultural milieu, only by having contact with all those who are now, and who in future will be, engaged in promoting spiritual culture and knowledge.[78]

These remarks speak for themselves. They would be wise remarks regarding historical context even if they had been written by an intellectual historian. However, given the context, i.e., the fact they were written by a great physicist reflecting on his field and on his own contributions makes them quite unique. Moreover, the perspective of the meta-theoretician influences his own theory:

> What has all this have to do with quantum jumps? I have been trying to produce a mood that makes one wonder what parts of contemporary science will still be of interest to others than historians 2,000 years hence...I am thinking of the theory of epicycles. I confess to the heretical view that their modern counterpart in physical theory are the quantum jumps.[79]

In the summer of 1960, shortly before his death, Schrödinger once again spent the summer in his beloved Alpbach. There he wrote the foreword to a small book, two essays actually published by Paul Zsolnay in 1961 with the title *Meine Weltansicht*.[80] The two essays, the 1925 "Suche nach dem Weg" (not published earlier) and the 1960 "Was ist wirklich?", can be seen as a kind of summary. Here I will I refer to the German originals, because with Schrödinger, as with Einstein, or Freud, it is extremely important to consider his own choice of words and to closely observe the nuances of the German version very often lost in translations. Other thinkers had a "Weltbild" (Einstein): a world view is an objectivist formulation. It is a subjective angle from which objective reality is viewed. "Weltansicht", on the other hand, is associated much more with a personal opinion ('meiner Ansicht nach...'= in my view) is a dialogue, a relativistic form of discourse. The "Suche nach dem Weg" is almost like a pun, combining both the descriptive and normative. The English "look for the way" is much more normative even though it retains some of the ambiguity. The "Was ist wirklich" (What is real?), however, is not a question. At least not in the real sense. I thus question the accuracy of its translation as "what is real?" Rather, it should read something like: "Let us sit down and discuss, what according to our respective opinions, is really the case with regard to the world." And this is precisely what his meta-theory deals with.

In the very short (less than two pages) "foreword" Schrödinger's often quoted heroes are mentioned, for example, Max Planck and Ernst Cassirer who were major philosophical influences. Hasenöhrl is once again cited as *the* teacher of physics who strongly influenced him. He also names other philosophers who were important for him: Spinoza, Schopenhauer, Mach, Richard Semon and Richard Avenarius.

The whole book deals with his philosophical ideas which have no direct connection with physics at first sight:

> Not a word is said here of a causality, wave mechanics, indeterminacy relations, complementarity, on expanding universe, continuous creation, etc...because I do not think that these things have as much connection as is currently supposed with a philosophical view of the world.[81]

Schrödinger sees Kant's impact as being the elimination of theoretical metaphysics. This is irreversible, and indeed science must be safeguarded from the encroachment of metaphysics. For him, metaphysics is initially a kind of uncertain idea, which then becomes physics. Yet metaphysics is an "indispensable basis of knowledge, in both a general and particular"[82] sense which serves as a scaffolding for our house of knowledge. The way metaphysics becomes part of science through "clarification and change in the philosophical point of view."[83] This change occurs also because words have their meaning only in a given context, and we can never really experience past thoughts as opinions. Following Cassirer, he traces all of man's philosophical quest to the Epicurean *wonder*; wonder at the phenomenon, the unexpected, that what is not *normal*! Yet Schrödinger believes we should not have been surprised since not knowing any different than our observable world, we should not have any particular expectation or anything 'normal' to compare to.

Schrödinger believes that the self (the psychological entity) and the external world (the object of physics) are made up of the same elements: sensations and perceptions. It is this basic monism, deeply rooted in Vedantic philosophy, according to which "the plurality we perceive is only an *appearance*; *it is not real.*"[84]
Schrödinger cites as sources Mach, Avenarius, Schuppe and the Upanishads:

The external world and consciousness are one and the same thing in so far as both are constituted by the same primitive elements. But we are then hardly ever using a different formula whether we express the essential *community* of these elements in all individuals by saying that there is only *one* external world or that there is only *one* consciousness.[85]

Schrödinger relies heavily on Semon's concept of *mneme*.[86] The 'mneme', the 'memory' of the organic and thus somewhat unconscious is the link Schrödinger needs to connect the conscious and unconscious in his absolute monism.

It is the peculiarity which Semon calls *mneme* by which a particular reaction, set in motion once, or more than once, by some stimulus-complex, gets 'drilled in' in such a way that in later similar occurrences only a part, and often a very small part, of the original stimulus-complex is needed in order to achieve the same result.[87]

This brings him then to the metaphysical conclusion:

Becoming is conscious, being unconscious.[88]

Language is the central mediator between the conscious and unconscious or between the self and the external world. Again it is an element shared by all, but it is the psychological aspect that interests Schrödinger.

...it is astonishing that, despite the absolute hermetic separation of my sphere of consciousness from all others (which no clear-thinking person denies) the origin and development of a common language, set in motion by the imitative instinct, as briefly sketched above, leads inevitably to the recognition of a far-reaching structural similarity between certain parts of our experience, the parts of which we call external; it can be expressed in the brief statement that we all live in the same world.[89]

In a recent article, the Italian physicist-philosopher Bruno Bertotti generalizes Schrödinger's attitude: he calls it r*ational mysticism* and detects it in Einstein and Eddington as well as in Schrödinger:

In the development of theoretical physics during the second quarter of this century one can recognize, I believe, the influence of a particular view of the world which could be called 'rational mysticism'...the 'existent' is a whole, and its unity can be apprehended intuitively; this apprehension goes together with a

deep feeling of wonder; the physical world can be understood rationally in terms of mathematical structures based upon the criterion of simplicity.
···Einstein...and Eddington...shared with Schrödinger a deep feeling of the unity of the existent.[90]

Does all of this give us a coherent philosophical picture of Schrödinger's "Weltansicht"? I am afraid that the answer is "not quite". Granted, we get a feeling for the kind of mind which is at work and for tendencies and motives. We grasp a broad 'culture-dependent, context-bound historical relativism. And above all we face a critical, yet sometimes enthusiastically non-critical, iconoclastic way of thinking. Clearly, Schrödinger was anti-positivist, something his friend and frequent opponent, Karl Popper, accused him of being.[91] In his biology he was a staunch Darwinian evolutionist but at the same time he tried to save as much of Lamarckism as he could.[92]

Schrödinger was also very critical of the scientific establishment and of the prevalent scientific style. What he resented was the positivistic, success-oriented absolutism which made, in his eyes, most scientists not open enough to new ideas and to contextual influences.

He repeatedly stressed that the knowing mind cannot study itself (a broad Gödel theorem!), but at the same he admitted that everything we know of the world is a construct of our mind, as already said above:

Die Welt ist ein Konstrukt aus unseren Empfindungen, Wahrnehmungen, Erinnerungen.[93]

The spirit, on the other hand, is free, and:

we shall not expect the natural sciences to give us direct insight into the nature of the spirit.[94]

NOTES

1 Preface to *Science and Humanism*, Cambridge 1952.
2 ibid., p. 4, quoting Plotinus: (Enn VI 414)
3 ibid., p. 4.
4 ibid., p. 8.
5 ibid., p. 53.
6 ibid., p. 60.
7 ibid.
8 ibid., p. 62.
9 ibid., p. 67, Cassirer is quoted from his *Determinismus und Indeterminismus in der modernen Physik*, Göteborg 1937.
10 "Nature," July 4, 1936, reprinted *Ges. Abh. 4* (2984) 364-5.
11 E. Schrödinger, *What is Life*?, Cambridge 1967, p. 93.
12 The lecture was held on February 18, 1932 at the Phys. Math. Klasse der Preussischen Akademie der Wissenschaften, and an augmented version published by Verlag Johann Ambrosius Barth in the same year, together with "Der Indeterminismus in der Physik" (held in 1931). The booklet is dedicated to Frank Exner. A somewhat misleading, almost free translation into English was done by James Murphy, who edited a volume of Schrödinger essays *(Science and Human Temperament)* in 1935. The essay in question here is divided into parts with two quite tendentious titles: "Is Science a Fashion of the Times" and "Physical Science and the Temper of the Age." Both are quite different from "milieubedingt".
13 Schrödinger is probably quoting from memory. In a study of Proudhon and Courbet, Zola says: "Une oeuvre d'art est un coin de la création vu a travers un tempérament." Mes Haines, 1866, in E. Zola, *Le Bon Combat*, Collection Savoir, Hermann 1974, p. 38.

14 The English version, from the 1957 Dover edition titled "Science Theory and Man," p. 86.
15 The German original from 1932 Barth pamphlet, op. cit., pp. 35-36.
16 ibid., pp. 37-38.
17 ibid., pp. 50-51.
18 *Ges. Abh. 3* (1984), p. 711.
19 This is not the context to go into detail. However, the above thesis is, in my opinion, the central theme in Klimt and Schiele, Schönberg and Mahler, Schnitzler and Karl Kraus, Mauthner and Wittgenstein.
20 One of the best comprehensive short papers is Dieter Flamm: "Boltzmann's Influence on Schrödinger" in C. U. Kilmister (ed.), *Schrödinger*, C.U.P. 1987, pp. 4-15. Flamm, who teaches at the Technische Universität in Vienna, is Boltzmann's grandson.
21 Franz Exner, "Über Gesetze in Naturwissenschaft und Humanistik." *Die Feierliche Inauguration des Rektors der Wiener Universität für das Studienjahr 1908/1909* (Vienna 1909), pp. 43-87, delivered on October 15, 1908.
22 Paul A. Hanle, "Indeterminacy before Heisenberg: The Case of Franz Exner and Erwin Schrödinger" HSPS (1979), pp. 225-269.
23 ibid., p. 226.
24 "The Statistical Law in Nature," *Nature 153* (1944), p. 704, reprinted in *Ges. Abh. 1* (1984) p. 452.
25 ibid., pp. 451-452.
26 Dieter Flamm, op.cit.
27 It is a thought-provoking fact that many great physicists, who have and proudly acknowledge a world-view, had *one* main teacher teach them the "whole" of physics! In his inaugural lecture in Berlin on July 4, 1929, Schrödinger, referring to Hasenöhrl's inaugural lecture in Vienna 1907, said: "Hasenöhrl...dem ich die Grundlage meiner wissenschaftlichen Persoenlichkeit danke." from "Antrittsrede des Herrn Schrödinger," Sitzungsberichte der Akademie der Wiss. Physikalisch-mathematische Klasse (1929), reprinted in: *Ges. Abh. 4* (1984), pp. 303-309. Parts of this speech are translated by James Murphy in his biographical introduction to the 1935 "Science and the Human Temperament," op. cit. There is a strange and telling mistranslation of the above quotation: "Hasenöhrl to whom I owe my whole scientific outlook." 'Scientific outlook' differs considerably from the 'basis' or 'foundations'of my scientific personality!
28 F. Exner, op. cit., quoted here from Paul Hanle, op. cit., p. 238 and his translation. The original reads: "Waeren wir im Stande die Molekularbewegung so ungeheuer zu verlangsamen, dass wir den einzelnen Vorgaengen folgen koennten, so wuerden wir nichts wahrnehmen als ein Chaos zufaelliger Ereignisse, indem wir vergeblich nach einer Gesetzmaessigkeit suchen wuerden." p. 55.
29 D. Flamm, op. cit., p. 10, quoting from E. Schrödinger, "Irreversibility," Proc. R. Irish A. *53* (1950), pp. 189-195, reprinted in: *Erwin Schrödinger, Gesammelte Abhandlungen in 4 Bde.*, Verlag der Oesterreichischen Akademie der Wissenschaften, Vienna 1984. *Ges. Abh. 1* (1984), pp. 485-491. The quotation is on page 191 in the original.
30 Exner, *Über Gesetze...*, op.cit., p. 50.
31 See my "A Programmatic Attempt towards an Anthropology of Knowledge," in: E. Mendelsohn and Y. Elkana (eds.), *Sciences and Cultures*, Reidel, 1981, pp. 1-76.
32 On this see the illuminating paper by Yemina Ben-Menachem, "Struggling with Causality: Schrödinger's Cat," *St. H. P. S.*, September 1989.
33 It is strange that Schrödinger did not contribute to the Schilpp volume on Einstein. Yet, in a 1929 article ("Einstein Explained") in a popular American periodical it was reported that Einstein himself suggested that Schrödinger be asked to interpret his theories. Einstein explained in "World's Work" (1929), pp. 52-56, reprinted in *Ges. Abh. 4* (1984), pp. 298-302.
34 "Might perhaps energy be a merely statistical concept?" Based on a lecture, delivered in Vienna on March 26, 1958 at a joint meeting of the Austrian Physical Society and the Chemical-Physical Society," as the editor of the *Ges. Abh. 1* (1984), p. 502 says.
35 Quoted by Hanle, op.cit. p. 236.
36 F. Exner, "Vorlesungen über die physikalische Grundlage der Naturwissenschaften", 1919, lectures 88 and 89. These lectures were much more widely read than his 1908 Rektoratsrede.

37 "Woher stammt nun der allgemein verbreitete Glaube an die absolute, kausale Determiniertheit des molekularen Geschehens...? Einfach aus der von Jahrtausenden erhaltenen Gewohnheit, kausal zu denken..." in: "Was ist ein Naturgesetz?" inaugural lecture at the University of Zurich, held on " December 9, 1922, repr. in a collection of essays bearing the same title, Munich 1987, p. 15.

38 ibid., the English translation appeared in: Science, Theory and Man, op.cit., p. 147.

39 ibid., p. 146.

40 Y. Elkana, "A Programmatic Attempt", op. cit.

41 E. S., "Anmerkungen zu Kausalproblemen - a letter to Reichenbach," published in *Erkenntnis* III (1932) 55-70 reprinted in *Ges. Abh. 4* (1984), 335-340.

42 More on that again, Hanle op. cit., and Hanle's doctoral dissertation. The exchange continued both in articles and in private correspondence. See: Bohr, Kramers, Slater: "The Quantum Theory of Radiation", E. S., "Bohr's neue Strahlungshypothese und der Energiesatz." *Die Naturwissenschaften 12* (1924), pp. 720-724. R. Stuewer, "The Compton Effect". E. S., to Bohr, May 24, 1924, *Bohr Sci. Corresp.* Microfilm 16, section 2 (Archive for History of Quantum Physics).

43 E. S., "Über das Comptoneffect," *Ann. d. Phys. 82* (1927).

44 E. S., "Might perhaps Energy be a merely Statistical Concept?" in: *Il Nuovo Cimento 9* (1958), pp. 162-70, repr. *Ges. Sch. 1* (1984), pp. 502-520.

45 Y. Elkana, "A Programmatic Attempt", op. cit.

46 E. S., "Anmerkungen zu Kausalproblemen - a letter to Reichenbach" published in *Erkenntnis* III (1932), pp. 55-70, reprinted in *Ges. Abh. 4* (1984), pp. 335 - 340.

47 "Antrittsrede des Herrn Schrödinger", *Sitzungsberichte der Preuss. Akad. d. Wiss. 4. Juli 1929*, reprinted in *Ges. Abh. 4* (1984), p. 304. Planck responded in his capacity as Secretary to the Academy, firmly representing the causal view of nature.

48 ibid.

49 "Ist die Naturwissenschaft milieubedingt?", op.cit., pp. 50-51.

50 Preface to *What is Life?* (based on lectures held in 1943), op. cit.

51 Linus Pauling, "Schrödinger's contribution to chemistry and biology" in C. W. Kilmister (ed.), *Schrödinger* C.U.P. (1987), p. 228.

52 M. F. Perutz, "Erwin Schrödinger's 'What is Life?' and Molecular Biology" in C. W. Kilmister (ed.) op.cit., p. 243 (he, too, analyzes Schrödinger's failure - to be taken up below). Against this background see a letter written by Francis Crick to Erwin Schrödinger on August 12, 1953: "Dear Professor Schrödinger, Watson and I were once discussing how we came to enter the field of molecular biology and we discovered that we had both been influenced by your little book *What is Life*? We thought you might be interested in the enclosed reprints - you will see that it looks as though your term 'aperiodic crystal' is going to be a very apt one." In: G. Kerber, A. Dieck, W. Kerber (eds.), *Erwin Schrödinger, Documents, etc.*, Fassbaender, Vienna 1987, p. 119.

53 *What is Life?*, preface, C.U.P. 1967, p. 1. Incidentally not to refer to this apology when making their attack is, to say the least, uncharitable on the part of Pauling and Perutz.

54 ibid., p. 5. Perutz, op. cit., does not want to understand the point and criticizes Schrödinger even for the term: "This leads him to suggest that the gene is a linear one-dimensional crystal, but lacking a periodic repeat: an aperiodic crystal. I find it strange that he does not call it a polymer, as Delbrück did." p. 241.

55 ibid., p. 5

56 *What is Life*?, op. cit., p. 5.

57 *What is Life*?, op. cit., p. 79.

58 ibid., p. 9.

59 "The Spirit of Science" in Eranos Yearbook: *Spirit and Nature*, Bollingen Series XXV (1954).

60 ibid., p. 324.

61 Festrede, gehalten bei der Eröffnung der fünften Weltkraftkonferenz, Wien 1956. Reprinted in *Ges. Abh. 4* (1984), p. 585.

62 ibid., p. 586.

63 Interviewed by J. W. N. Sullivan, "The Observer", Sunday, January 11, 1931, reprinted *Ges. Abh. 4* (1984), p. 334.

64 E. S., "Autobiographische Skizze," (Les Prix Nobel 1933, pp. 86-88), Stockholm, Norstedt and Soener, 1935, reprinted *Ges. Abh. 4* (1984), pp. 361-363.
65 *Science and Humanism*, op. cit., p. 53.
66 *Science and Humanism: Physics in our Time*, op. cit., pp. 40-41.
67 ibid., p. 5.
68 ibid., p. 608.
69 op. cit.
70 "Ist die Naturwissenschaft milieu-bedingt?" op. cit., p. 32.
71 ibid., p. 42.
72 "Conceptual Models in Physics and Their Philosophical Value" (lecture delivered to the Physical Society of Frankfurt, December 8, 1928, reprinted in: *Science, Theory and Man*, Dover 1957, p. 160; reprinted in *Ges. Abh. 4* (1984), p. 288).
73 ibid., p. 164-165.
74 *Nature and the Greeks*, Cambridge University Press, 1954. These were the Sherman lectures Schrödinger held at the University College London on May 24, 26 and 31, 1948.
75 ibid., p. 1.
76 "Are there Quantum Jumps?," in: *BJPS 3* (1952) 3-17; pp. 19-28, reprinted in *Ges. Abh. 4* (1984), pp. 478-502.
77 ibid., pp. 3-4.
78 ibid., pp. 19-20.
79 ibid., p. 6.
80 Published by C.U.P. in English as *My View of the World* and the two parts as "Seek for the road" and "Which is real?", op. cit.
81 ibid., p. VIII.
82 ibid.
83 ibid.
84 ibid., p. 18.
85 ibid., p. 37.
86 R. Semon, *Die Mneme*, Leipzig, 1904 and *Die Menschlichen Empfindingen*, Leipzig, 1909.
87 ibid., p. 44.
88 ibid., p. 51.
89 ibid., p. 105. The inherent tension between self and the world, the real and the thought-world, linked by language is as true in physics as it is in philosophy. Discussing "reality" of matter-waves in his 1928 "4 Lectures on Wave Mechanics" held at the Royal Institution in London on the 5th, 7th, 12th and 14th of March 1928 and published in London in 1928, he says:
"The statement that what *really* happens is correctly described by describing a wave-motion does not necessarily mean exactly the same thing as: what really *exists* is the wave-motion. We shall see later on that in generalizing it as an *arbitrary* mechanical system we are led to describe what really happens in such a system by a wave-motion in the generalized space of its coordinates *(q-space)*. Though the latter has quite a definite physical meaning, it cannot very well be said to "exist;" hence a wave-motion in the space cannot be said to "exist" in the ordinary sense of the word either. It is merely an adequate mathematical description of what happens."
90 Bruno Bertotti, "The Later Work of E. Schrödinger," *S. H. P. Sci. 16* (1985), p. 83.
91 K. R. Popper, *Unended Quest*, p. 135. Our relations had been somewhat stormy. Nobody who knew him will be surprised at this. We disagreed violently on many things. Originally, I had taken it almost for granted that he, with his admiration for Boltzmann, would not hold a positivist epistemology, but our most violent clash was sparked off when I criticized one day (around 1954 or 1955) the Machian view now usually called "neutral monism" - even though we both agreed that, contrary to Mach's intentions, this doctrine was a form of idealism.
92 "Unhappily Lamarckism is untenable. The fundamental assumption on which it rests, viz. that our acquired properties can be inherited, is wrong. To the best of our knowledge they are not. The single steps of evolution are those spontaneous and fortuitous mutations which have nothing to do with the behaviour of the individual during its life-time. And so we appear to be thrown back on the

gloomy aspect of Darwinism that I have depicted above. I now wish to show you that this is not quite so. Without changing anything in the basic assumptions of Darwinism, we can see that the behaviour of the individual, the way it makes use of its innate faculties, plays a relevant part, nay, play the most relevant part in evolution. There is a very true kernel in Lamarck's view, namely that there is an irrescindable causal connexion between the functioning, the actuality being put to profitable use, of a character - an organ, any property or ability or bodily feature - and its being developed in the course of generations, and gradually improved for the purposes for which it is profitably used. This connexion, I say, between being used and being improved was a very correct cognition of Lamarck's, and it subsists in our present Darwinistic outlook, but it is easily overlooked on viewing Darwinism superficially. The course of events is almost the same *as if* Lamarckism were right, only the "mechanism" by which things happen is more complicated than Lamarck thought. The point is not quite easy to explain or grasp, and so it may be useful to summarize the results in advance. To avoid vagueness, let us imagine an *organ*, though the feature in question might be any property, habit, device, behaviour, or even any small addition to, or modification of, such a feature. Lamarck thought that the organ (a) is used, (b) is thus improved and (c) the improvement is transmitted to the offspring. This is wrong. We have to think that the organ (a) undergoes chance variations, (b) *the profitably used ones* are accumulated or at least accentuated by selection, (c) this continues from generation to generation, the selected mutations constituting a lasting improvement. The most striking *simulation* of Lamarckism occurs - according to Julian Huxley - when the initial variations that inaugurate the process are not true mutations, not yet of the inheritable type. Yet, if profitably, they may be accentuated by what he calls *organic selection*, and, so to speak, pave the way for true mutations to be immediately seized upon when they happen to turn up in the "desirable" direction." In the 1950 BBC talks: "The Future of Understanding," reprinted in *Mind and Matter* and in *Ges. Abh. 4* (1984), p. 469.

93 Festrede, held at the opening of the fifth 'Weltkonferenz' in Vienna 1956, op. cit.

94 'The Spirit of Science' in Eranos Yearbook: op.cit.

ERWIN SCHRÖDINGER'S POSITION IN THE EINSTEIN-BOHR DEBATE

ULRICH RÖSEBERG
Institut für Philosophie
Otto-Nuschke-Straße 10/11, 1086 Berlin, Germany

Albert Einstein and Niels Bohr belong to the pioneers of quantum physics. Their contribution to the development of the theory was based on different expectations and research programs. After non-relativistic quantum mechanics had been developed, their debate touched on all the central aspects of the interpretation of this theory, which revolutionized the world view of physics. They were thus also confronted with the epistemological, methodological and world-view related implications of the development of physics in the 20th century. Therefore, it certainly does no injustice to Erwin Schrödinger's extraordinary scientific contributions to the development of wave mechanics to analyze his position in the intellectual debate between those two scholars, who admired each other and later became friends.[1] We will focus here on letters, or excerpts of letters.

Einstein and Bohr both summed up and evaluated the entire debate, each from his respective point of view.[2] The pertinent secondary literature has meanwhile become so vast that it is almost impossible to survey.[3]

1. SCHRÖDINGER'S PHYSICAL AND PHILOSOPHICAL IDEAS BEFORE THE FORMULATION OF WAVE MECHANICS

Werner Heisenberg's first breakthrough and the contributions he elaborated together with Max Born and Pascal Jordan as well as P. A. M. Dirac's work evolved from the research program drafted by Bohr in 1912/13, and later elaborated together with Sommerfeld. This program did not, however, form the basis of Erwin Schrödinger's scientific contributions to the non-relativistic quantum mechanics emerging at the time. In a systematic retrospective, however, Schrödinger's wave mechanics can be interpreted as both resulting from and transcending Bohr's original research program, even though this may in some respects be contrary to what Schrödinger intended.

The main purpose of this paper is to outline some of the physical notions, methodological principles and philosophical convictions on the basis of which Schrödinger began to elaborate wave mechanics in late autumn of 1925 after Einstein had drawn his attention to de Broglie's doctoral thesis. It will become apparent that these ideas were taken up time and again and elaborated in discussions and interpretations beginning in 1926.

Schrödinger's reservations against the Bohr-Sommerfeld research program in atomic physics are probably mainly due to the fact that he - like Einstein - could and would not accept the inconsistencies and ad hoc assumptions of the model theory which was repeatedly adapted to the

J. Götschl (ed.), Erwin Schrödinger's World View, 135–152.

growing amount of experimental evidence. What he found lacking for a long time was basic theoretical framework in which the emerging problems could be solved step by step on the basis of consistent principles. He succeeded in overcoming precisely this desideratum with his wave theory, whereas Einstein had tried in vain to achieve a breakthrough in this direction.

Hence, although Schrödinger had from the very beginning been closer to Einstein than to Bohr in methodological terms, we cannot discard the fact that in one decisive point of the development of ideas in theoretical physics he did not side with Einstein. Like Planck - and Bohr as well - he did not agree with Einstein's light quantum hypothesis. From Maxwell's electrodynamics, Schrödinger inferred that light can only propagate in the form of waves. Moreover, his conviction that there is a continuity in natural phenomena seems to have become firmly rooted in a philosophy of nature at an early date.

In his manuscripts on philosophy that were published in the volume *Suche nach dem Weg* (Search for the Way) in 1925 (most of them written in 1918/19, one while at grammar school and another in 1925), Schrödinger deals, last but not least, with ancient Indian philosophy. Being a physicist searching for a philosophical way, he was then fascinated by the perspectives of the Vedic fundamental conviction that "cognizing, feeling and volition are essentially eternal and immutable and ... numerically only *one* thing in all humans, even in all feeling beings." When delimiting this from mechanistic notions about the part and the whole, as well as from the pantheism of Spinoza, he added by way of explanation:

> you - and likewise every other conscious being taken by itself - are all in all. Therefore this life of yours that you live is also not a piece of universal events, but the *whole* in a certain sense. Only this whole is not so that it can be surveyed at *one* glance. - It is well known that this is what the Brahmans express in the holy, mystical, properly speaking, so simple and clear formula: *Tat twam asi* (this is you). - Or also in words like: I am in the east and in the west, below and above, I *am this entire world*.[4]

Schrödinger modified this thought so that it could be relatively easily reconciled with modern natural scientific thinking by saying that "the acts of propagation through which a series of ancestors emerge from one another in a successive manner constitute no proper interruption, but rather only an interlocking of both corporeal and mental life."[5] The reader can easily find a number of such passages. Could it not be that one of the sources of Schrödinger's physical approach to the quantum problem is to be found in these philosophical ideas in which continuity is the central notion?

It is virtually impossible today to answer this question in a definitive way. However, if it is raised here, it is meant to discourage light-hearted attempts to interpret Schrödinger's later philosophical statements, as well as his attempts to develop a specific physical view of the world as the scurrilous ideas of the aging scholar. In this respect, the intellectual biography of Schrödinger reveals much more continuity than some people may be willing to admit.

On another point, however, i.e., the important epistemological issue of subject-object, this scholar contradicts himself in a way one would hardly expect from a man of his scientific calibre. Regarding our knowledge of nature, Schrödinger advocates from the very beginning a strict distinction between subject and object. Later on he opposes all attempts to overcome this distinction methodologically. In all other spheres, however, he tries to transform this distinction into an identity, compensating, as it were, for the concomitant hardships, which lead him to mysticism. Again the intellectual points of departure lie in the period before the formulation of wave mechanics.

Perhaps what Max Born once said about himself also applies to Erwin Schrödinger:

I am a logician only on a 'small scale,' when I have to work out a clear problem. 'On a major scale,' where essential issues are concerned, the application of strict logical methods seems absurd to me - there are always premises which we do not know or are unaware of.[6]

Schrödinger himself asserted: "A purely intellectual view of the world without any mystics is impossible."[7] In one of his reflections, dating from 1925, he tried to delineate mystics from cognition:

.. the expression of a thought in a *word* that is easy to convey and to retain is like the work of a silk worm. Only giving shape to the material makes it valuable. But in the light of day it becomes rigid and something strange which can no longer be molded. Indeed, we can thus recall the same thought more easily and at will, but perhaps we can never again experience it as naturally as before. - Therefore the ultimate and most profound insights are and always will remain voce meliora (better than voice).[8]

2. FIRST REACTIONS TO WAVE MECHANICS (1926-1927)

The young physicists from Göttingen (most notably, Heisenberg and Jordan), who had just broken new ground in the new field of quantum mechanics with their matrix mechanics, strongly opposed Schrödinger's wave mechanics. On the other hand, Bohr, Einstein, Planck, Lorentz, and also Pauli, were deeply impressed by it from the very beginning. One of Bohr's first statements dating from late January 1926 was as follows:

Even if wave mechanics cannot help us with the problem of nature's discontinuity, it is a most beautiful discovery.[9]

He only wrote to Schrödinger on September 11, 1926 to invite him to a discussion to Copenhagen which was to be attended by Pauli as well. His comment on wave mechanics in early February: "Perhaps his way is not so mad."[10]

A short time later, independently of Schrödinger, Pauli found the proof for the mathematical equivalence of both representations of quantum mechanics.[11]
Planck wrote that he read Schrödinger's treatise "like a curious child who listens with great interest to the solution of a riddle which it had been trying to solve for a long time."[12] Lorentz voiced his appreciation, but was still undecided. On Einstein's first inquiry into the matter he responded by arguing in favor of Schrödinger's approach. In late April, a letter from Berlin arrived in Zurich which read:

I am convinced that by your formulation of the quantum condition you have made a decisive advance, as much as I am convinced that the way of Heisenberg and Born is misleading.[13]

The recollections of the participants of Schrödinger's Copenhagen discussions with Bohr in October 1926 are very informative.[14] A letter written by H. A. Kramers also draws a good picture of the central issues of the discussion. At the time Kramers had just completed his assistantship with Bohr, with whom he had worked for several years. Pointing out the potential of Schrödinger's theory, the latter writes:

I hope that the strange mixture of formulae referring to charged dot-like masses and such formulae related to oscillation phenomena will soon to lose their mystical tinge. The dissolution of the antithesis light waves and light quanta will probably proceed in parallel with this. Perhaps the decisive step towards this dissolution has yet to be made. The atoms in their theory, as far as it known to us, as well as in Heisenberg's formulation, are at first only spectre atoms. As far as I can see they only provide a descriptive background for the migration of energy as expressed in the measurements.[15]

Despite the meanwhile existing proof of the mathematical equivalence of Heisenberg's matrix mechanics and Schrödinger's wave mechanics, there was still the important issue, which theoretical approach to prefer in physical terms. Moreover, it was necessary to determine how to interpret for physics the mathematical formalism of the new theory, as well as what this theory states about the objective regularities of the range of objects for which it was drafted. Schrödinger's position on this issue is revealed in a letter that he wrote to M. Born in Göttingen after returning from Copenhagen. In this letter he asks whether:

the existence of "certain discontinuities in nature" is really to be considered already proven,

and adds the following question:

Have you considered the fact that all these experiments were begun under the powerful auspices of exactly that theory, and have never been regarded from another viewpoint than yours, that is to say, the theory which puts such 'discontinuities' somewhat dogmatically at the beginning of all considerations?

In any case, Schrödinger suggested "searching for facts with the greatest care and possibly for years, if one cannot avoid dogmatically introducing these discontinuities into the theory as something principle."

In this context, his deliberations were, at least in 1926, by no means to be understood to the effect;

...that I insist on sticking to my points. Nothing less than this. At least for me, the first advance was so difficult that I am not worried at all, if my explanation must be changed.[13]

On his return from Copenhagen, Schrödinger wrote to Bohr expressing his gratitude for the hospitality which had meanwhile become proverbial. In this letter he described how he felt after the strenuous disputes with the seemingly indefatigable interlocutor as follows:

It is possible that the stubbornness, with which we conducted our dialogues and with which I continued to adhere to my 'wishes' for a physics of the future, may have left you in the end with the impression that the general and specific objections that you raised against my views had not made any real impression on me. That is certainly not the case. In a certain sense I can say: the psychological effect of these objections - in particular the numerous specific cases in which for the time being my views apparently can hardly be reconciled with experience - is probably even greater for me than for you. And this for the very reason that you, as it seems to me, have found a certain preliminary position in the view that all the apparently perceivable representations are really only to be seen symbolically, above all - as pointed out by Dirac in his most recent publication, and also advocated by Born - that the c's, the amplitudes or coefficients of the individual proper vibrations, merely convey statistical statements on the behavior of a large number of identical systems, and do not describe the behaviour of a single system. However, I am quite unable to set my mind at rest with this preliminary solution. It appears to me in *general* just as inapplicable as my own... What I vaguely envision is only the idea: even if a hundred attempts have failed, one ought not to give up hope of arriving at the goal, I don't say through classical pictures, but by means of logically consistent conceptions of the true nature of the space-time events. It is extremely likely that this is possible.[17]

And this is exactly what Bohr already contested at that time with great determination.

Above all, it must be noted that Schrödinger did not oppose the probabilistic interpretation of the wave function because this interpretation granted chance a place which it did not have in mechanical materialism with determinism.[18] Schrödinger's scepticism with regard to this issue was at first related to the fact that - in Bohr's work even more than in Born's - the reference to statistics was already very closely linked with the changes in the truth concept and thus with the comprehension of theory before Heisenberg's uncertainty principle. In order to minimize as much as possible the difficulties emerging from the rejection of the microphysical pictures of space and time, Bohr then spoke about the symbolic nature of the quantum mechanical formalism and the corresponding signs, without clearly indicating what these symbols really stood for. The transition from formalism to observation and measurement remained unclear for the time being. Schrödinger was pressing for more clarity: for him, this involved a space-time picture of the individual microphysical events, which Bohr already saw as impossible.

A letter written by Schrödinger in that period reads for instance:

Bohr's attitude to the atomic problems...is really strange. He is completely convinced that an understanding in the usual sense of the word is impossible. In this context, the conversation almost always immediately drifts towards philosophical issues, and soon one no longer knows whether one really adopts the attitude he is opposing or whether one really must oppose the attitude *he* adopts. Quite certainly, the use of discriptive pictures as practiced by de Broglie and myself has not been carried out far enough, so as to account even for the most important facts. And it is just as probable that (we) here are on the wrong track...[19]

Born evaluated Schrödinger's endeavors to concentrate further theoretical development on the wave aspect of quantum mechanics, as an attempt to adapt the new theory to the standards of classical physics. In 1926, for a short time, he also shared the hope that the fundamental physical problems which could not initially be solved by Heisenberg's formalism could be dealt with relatively easily by means of Schrödinger's wave mechanics. The discussion of the impulse problems, which is known to have led to a probabilistic interpretation of Schrödinger's wave function, rapidly brought Born into the camp of his younger co-workers W. Heisenberg and P. Jordan as far as the debate about the superior physical representation of quantum mechanics was concerned. This, however, implied opposition to Schrödinger, Einstein and von Laue. Born now stated that Schrödinger's method had indeed led to great advances in mathematics but that the matrix view seemed superior in terms of principles.[20] The following statement by Born surprised and annoyed Schrödinger:

Among my younger co-workers a combative mood prevails, since the overwhelming triumph of wave mechanics seems to us to be unjustified by facts, and can be explained by meeting the demands of most people to return to classical forms of physics.[21]

On this issue the physicists from Göttingen were even more radical than Bohr. Born, however, emphasized time and again that in principal issues, he always tried to orientate himself after Bohr's philosophical ideas. This can be seen, for instance, in a letter written by Born that arrived in Zurich soon after Schrödinger's return from Copenhagen in autumn of 1926:

It seems to me that Bohr's way of speaking is the natural description and summary of a major domain of facts; therefore it is a task of any subtler theory to justify this way of speaking...[22]

Precisely this was the objective of Born's interpretation of Schrödinger's wave function. This may well be the main reason why it was not recognized as a particular achievement in Copenhagen. To Bohr it must indeed have appeared as some justification of his own way of speaking - Schrödinger, however, harbored objections from the very beginning, as "the philosophical aspect of it...is quite horrible."[23] He argued:

The entire construction of the radiating dipole moments which leads to such good results is completely transformed along the lines of Vaihinger, if one follows Born. Indeed, only one vibration is activated in each case, but 'everything is as if...' several were activated at the same time.[24]

This constantly recurring 'as if...' reminiscent of Vaihinger's philosophy had already been criticized by Schrödinger in his discussions with Bohr. However, at the same time he saw that Born's interpretation at once eliminated many difficulties, stressing that "until today there is not the shade of a reasonable approach for a different...calculation".[25]

In late November 1925, Schrödinger admitted to Bohr:

All in all, I am more unclear about the fundamental questions than ever before. That is a wonderful condition just at the moment when one is to travel to America to expound the wisdom which one does not possess before an amazed crowd.[26]

He wrote to Kramers:

At the moment I am again more uncertain than ever as to how to understand the whole matter. Born wants to interpret the 'c', the amplitudes of natural ocillation in terms of statistics, as the description of an ensemble. Originally, I wanted to grasp it as a description of the state of the individual systems. And then it seems to be like this: in certain cases one is compelled to take the one view, and in certain cases, the other view. This is very unsatisfactory.[27]

After Schrödinger had dealt in greater detail with Born's probabilistic interpretation of the wave function, he presented another serious objection in 1927. He compared his wave equation with Fokker's equation and underlined that the former is reversible, whereas the latter does not remain invariant, if the time direction is reversed. Therefore, Schrödinger concluded, the probability approach of quantum mechanics was still insufficient for describing irreversible events.

Hence, *although* one has assumed an 'absolute, primary' probability of the elementary event, one has to come up with a 'disorder hypothesis' at some point.[28]

But one month earlier he had written:

At present I think it is more important to work on details than to reflect on questions of interpretation, especially if they tend to be philosophical.[29]

However, these reflections on aspects of interpretation had entered a decisive stage in 1927 in view of Heisenberg's uncertainty relations and Bohr's ensuing complementarity argument. This is evidenced by the discussions on Bohr's paper presented at Como and the public commencement of the intellectual debate between Einstein and Bohr at the 5th Solvay Conference in Brussels.

3. BOHR'S COMPLEMENTARITY ARGUMENT AND SCHRÖDINGER'S CALL FOR A NEW CONCEPTUALIZATION (1927 - 1935)

Einstein received word from Copenhagen about the turn of events brought about by Heisenberg's uncertainty relations. In the pertinent letter Bohr points out the conceptual difficulties of quantum theory:

> It has, of course, long been recognized how intimately the difficulties of quantum theory are related to the concepts, or rather to the words that are used in an ordinary description of nature, which all have their origin in classical theories. These concepts only leave us with a choice between Charybdis and Scylla, depending on whether we direct our attention towards the continuous or the discontinuous aspect of the description. Yet, at the same time we feel that is the hopes, conditioned by our own customs, that lead us into temptation, insasmuch as it has up to now been possible to keep ourselves afloat among realities, as long as we are prepared to sacrifice any accustomed desire. This very fact that the limitations of our concepts coincide so closely with the limitations in our possibilities of observation enables us - as Heisenberg emphasizes - to avoid contradictions.[30]

As we know today, the formulation of the coincidence between the "limitations in our concepts with the limitations of our possibilities of observation" was not made by Heisenberg. It was, rather, one of the first tentative attempts to put forth Bohr's complementarity argument. In 1927, Niels Bohr prepared an overview of the current situation in quantum mechanics for the Volta Congress which was to take place at Como. The paper, which was prepared in permanent discussion with his co-workers, contained the outlines of what today is often called the "Copenhagen interpretation of quantum mechanics". This comprises Born's interpretation of the wave function, Heisenberg's uncertainty relations and Bohr's complementarity argument.

According to Bohr, *complementarity* establishes for quantum mechanics, a connection between pairs of concepts that are employed either simultaneously or alternatively in classical physics. The complementarity argument states that the conditions for observing a quantity or a concept and the possibilities for defining the appropriate other quantity or appropriate other concept limit and determine each other. Thus, complementarity implies far more than a dualism characterized by "both - and" formulations. The practical observation of the wave aspect of matter limits the possibility of speaking simultaneously about the particle aspect of matter and vice versa. According to Bohr, Heisenberg's uncertainty relations imply that with the determination of a quantity performed in the practical measurement, its canonically conjugated quantity (in correspondence with the relationship that can be indicated in mathematical terms) is limited in terms of possible definitions. Therefore, it no longer makes sense to speak about place and momentum, and to pretend that the only change is that they can no longer be measured simultaneously. The observation conditions for the determination of position and the definition possibilities for the determination of momentum thus limit and determine each other at the same time, in correspondence with Heisenberg's uncertainty relations.

Complementarity is not a philosophical demand quantum mechanics was forced to meet from the outside, but a result of theoretical development. In this respect, the development of physics highlights the fact that ontology and epistemology belong together. Quantum mechanics excludes an ontological manner of speaking, which is still feasible and unproblematic in classical physics. This is at least how Bohr's philosophical message was understood, which, of course, was not expressed in such clarity at the very beginning. Up to the present day, even Bohr's adherents do not all interpret his thinking in the same way. To be sure, the differences in interpretation are much greater among those who reject the core of Bohr's philosophy more or less stringently. Here I would like to point out that Schrödinger never conceived of Bohr's complementarity argument in this form, whereas Einstein approached but finally rejected this interpretation as unsatisfactory.

With regard to the first discussions on the complementarity argument, I would like to draw the reader's attention once more to Bohr's description of his outlook ("to keep oneself afloat among the realities") and his readiness "to sacrifice any accustomed concept." Such statements thoroughly contradicted the ontological and epistemological considerations on natural science that one finds in Einstein, Schrödinger and von Laue. They certainly recognized the fact that it was becoming increasingly difficult for physics as well, to define a reality concept which could be advocated in scientific terms. Scholars who accepted model theories only as a means to construct theories based on principles (and possibly on didactic grounds for the purpose of illustration) were, however, hardly willing to respond to this difficulty by "keeping oneself afloat among the realities". On the other hand, they showed more readiness "to sacrifice any accustomed concept" in view of new experimental evidence or theoretical constraints. They, too, were ready to do that, if it proved necessary.

Whereas Schrödinger's lecture at the Solvay Conference in 1927 triggered debates on some details of his presentation of wave mechanics, Bohr's paper kindled lively discussions about the fundamentals of gnoseological problems, in which, apart from Einstein and other scientists, also Born, Dirac, Heisenberg, Kramers, Schrödinger and Ehrenfest[31] took part. In 1928 Bohr published a thoroughly revised version of the lecture he had given in Como and in Brussels in 1927 in "Naturwissenschaften", providing new fuel for the controversy over the fundamental problems of quantum mechanics. Schrödinger became acquainted with this article only once the galley proofs were available and made it the topic of a speech he held at the Berlin Wednesday colloquium. He communicated the ideas expressed in this speech to Bohr to make sure that he had understood him correctly. In this letter he explicitly expounds for the first time an idea that is to determine Schrödinger's thinking in the years to come. He interpreted Heisenberg's uncertainty relations (he himself speaks of the "inaccuracy relation") as a limitation to the applicability of the *old* concepts of 'experience'.

> It now seems imperative for *new* concepts to be introduced for which this limitation *no longer* applies. Because what is in principle unobservable should not be included in our conceptual scheme and it should not be possible to represent it within the latter either. In an *adequate* conceptual scheme it should no longer appear as if our possibilities of experience were limited by unfavorable circumstances. However, it will no doubt be *very* difficult to invent this new conceptual scheme, since - as you emphasize so impressively - the refurbishing touches the deepest levels of our experience: space, time and causality.[32]

Bohr, however, wholeheartedly disagreed with the demand for "*new concepts*", as his complementarity argument was meant to show why in the new fields of experience old concepts must be preserved and how they must be modified at the same time. In his amiable manner, he clothed his protest in the diplomatic formulation that he was not quite in agreement.
He justified this as follows:

> Still I do not quite agree with the stress you place on the necessity to develop 'new' concepts. We have not only, as far as I can see, no basis for such a new-fashioning as it stands, but also the 'old' empirical concepts appear to me to be inseparably linked to the foundations of human means of visualization. True enough, the apparent contrast between the superposition principle and the individuality postulate has

revealed the complementary nature of the space-time coordinates and conservation laws. But I think that we are concerned here with a philosophically consistent and hence satisfying extension, of the foundations of our description of nature. In my opinion it is also not an issue of a more or less arbitrary limitation to the applicablity of classical concepts. Rather, we have to recognize an inevitable feature of complementarity that emerges in an analysis of the concept of observation and which in many respects recalls the recognition of the general (feature of) relativity. Of course we do not possess in quantum theory, technical means comparable with that of the theory of relativity. However, I believe that also in this respect, quantum theory is approaching a certain temporary completion. Indeed, I believe, that it is already possible to say that any application of classical concepts that can be defined in unambiguous terms, may also be ascribed a physical interpretation.[33]

Einstein, who was informed about Bohr's letter, communicated his view to the addressee in unmistakable terms:

Your claim that the concepts p, q will have to be given up, if they can only claim such a "shaky" meaning, seems to me to be fully justified. The Heisenberg-Bohr tranquilizing philosophy - or religion? - is so delicately contrived that, for the time being, it provides a gentle pillow for the true believer from which he cannot very easily be aroused. So let him lie there. (May 31., 1928)[34]

Einstein, as we know, first directed his attack of the thesis of the completeness of quantum mechanics at the universality of Heisenberg's uncertainty relations. He wanted to show that these relations can also be violated within the theory's range of validity - an attempt that obviously failed because of Bohr's counterarguments. His words "let him lie there" thus reflected this experience.

However, Einstein and Schrödinger did not stick to this motto at all. This is evidenced by the debate between Einstein and Bohr at the Solvay Conference in 1930 as well as by Einstein's and Schrödinger's fundamental article from 1935 and a number of later statements. For a long time Einstein and Schrödinger seemed to be following a concerted strategy. Although they took up different issues (Einstein - "physical reality" in microphysics and observer; Schrödinger - applicability of quantum mechanics on the macrophysical level and observer), until 1935 (Schrödinger then focussed exclusively on problems of relativistic quantum field theories) their criticism basically followed the same lines. Paradoxes existing then within quantum mechanics prompted calls for a reformulation of the theory and for its conceptual expansion. The Berlin theorists became the antipodes of the Copenhagen theorists. Bohr usually remarked that the paradoxes indicated by Einstein and Schrödinger corresponded to the nature of quantum mechanics, and could also be mastered by means of the complementarity argument without any conceptual innovations.

When after Hitler's seizure of power in 1933, Germany's cultural and intellectual life was maimed beyound recognition by Nazi racial legislation and the relentless suppression of all forces that appeared dangerous to Nazism, Albert Einstein and Erwin Schrödinger were among the first scientists who emigrated. Thus the fundamental discussions on quantum mechanics could no longer be localized as polarizations between Copenhagen and Berlin.

The theorists, Planck, Einstein, von Laue and Schödinger, who worked together in Berlin until 1933, had much in common in their basic philosophical outlook, in their understanding of physical theories and theoretical development, as well as in their research. On all these levels, they differed from their junior fellow scientists who had then gathered in Copenhagen around Bohr. Some of the letters Schrödinger wrote in Berlin provide interesting insights into what led up to this shift of priorities within the international network of the leading physical research centers which was already foreseeable before 1933. For instance one of the letters he wrote to Bohr in 1930 reads:

I would like...to say quite explicity that *Berlin* is not what I would call 'the right place and the right food for thought' - unfortunately, but I cannot change it. As you know, Planck had no talent at all for 'setting up a school', von Laue has little talent, and I myself have about as little as Planck. Besides that, the whole operation is too big, one has to deal with too many people and with too much, and has too little time to devote to the individual.[35]

Here he did not even mention Einstein!!! Suffice it to say, he fitted very well into the picture drawn by Schrödinger. A few months later Schrödinger wrote:

We in Berlin are solitary persons, each too much pursuing our own private ideas, hardly stimulating young people and criticizing them too little. Whoever is very good himself, as for instance London and Neumann, feels free and good here. But we are not particularly good at showing people not so sure of themselves the right track and giving them the necessary guidance.[36]

He did not just complain about methodological aspects but also about the general state of the field, i.e., "about the presently quite depressing state of quantum mechanics." On another occasion Schrödinger says:

Honestly, now I know, after believing to have finally understood the development of recent years, that I am at the end of my tether. Recently, everything has gone wrong. A major new idea has to appear. Otherwise, despite all successes, we shall be left in a lurch. In quantum mechanics, we have progessed almost exactly as far as in electrodynamics before Faraday and Maxwell. That means, we have a marvellously beautiful remote action theory and cannot really go beyond that.[37]

At that time Einstein expressed a very similar view.

Today we have to admit that the major new breakthrough in this field that was expected and hoped for has yet to emerge. Therefore, we must ask whether such an idea is really necessary, or whether the Copenhagen interpretation is basically sufficient with all its philosophical implications, as first put forth by Bohr, and whether further cognitive advance will follow the shift of focus to deeper levels in the structure of matter.

Whereas Schrödinger became increasingly suspicious of the particle concept, Bohr emphasized that "all previous applications and successes of quantum theory"[38] are based on the assumption that elementary particles exist. In 1931, he agreed with Schrödinger that the previous viewpoints and methods had reached the limits of their capacity, and expressed the following hope:

To progress further, it may be necessary to develop new concepts. Accordingly, the existence of the particles and of the quantum of action appear from the very beginning by necessity, to imply inevitable a general principle.[39]

In 1935 the discussions on quantum mechanics reached a certain climax and a temporary completion. In that year an article written by A. Einstein, B. Podolsky and N. Rosen ("Can Quantum-Mechanical Description of Physical Reality be Considered Complete?") was published, as well as Bohr's response under the same title and Schrödinger's essay "Die gegenwärtige Situation in der Quantenmechanik" (The Present Situation in Quantum Mechanics).

Referring to Einstein and von Laue, Heisenberg stated:

Perhaps the understanding of quantum theory is in principle, complementary to an understanding of the psychology of those who do not want to understand quantum theory.[40]

Bohr found much that was beautiful about Schrödinger's article, but remarked that he could "understand his entire outlook as little as that of Einstein." [41] - Einstein and in particular Schrödinger had each expressed the hope that the problems that were still unsolved in relativistic quantum mechanics might cast new light on the controversial principal issues of non-relativistic theory. Meanwhile, Bohr could no longer share this hope; he proceeded from the assummption that for a deeper understanding of the epistemological implications of non-relativistic quantum mechanics no new theoretical approaches were needed.

In March 1936, Einstein had another meeting with Bohr in London and reported how this scientist:

in his nice friendly way said several times that he found it 'appalling' - even 'high treason' - that people like Laue and myself, but especially somebody like you wanted quantum theory to be pitted against the well-known paradoxical situation which was so drawn into the matter, was so demanded by the experiment. It was presumptuous to demand that nature have to comply with our preconceived views of 'reality'; we had modestly to note how things really happen in nature and were requested to adapt our thoughts to that. Such a profound inner conviction from such an extraordinarily clever man that I find it difficult to adhere firmly to my convictions, notably if I have none other than this: folks, there is something fishy here!

But I do think it is good if those who find this continue to violently press those others in all friendship, be it also to bring to light in the course of time an ever more lucid expression of Bohr-Heisenberg's viewpoint. Because the preceding expressions of opinion are still tortuous and Sybillinic for me. I really do not care who 'is right' in the end. However, I cannot endorse an idea which I neither can think clearly myself nor explain to a young person asking me to.[42]

In one of the last letters Einstein wrote to Schrödinger before the outbreak of World War 2, it becomes clear that meanwhile, the views held by the critics of the Copenhagen interpretation of quantum mechanics also differed considerably. Einstein considered the wave representation of matter to be incomplete and wanted to prove this with the help of Schrödinger's cat, by distinguishing between the two following options which are ultimately both rejected:

If one tries to understand the function as the complete description of a state (independently of observation), then this means that at the same time the cat is considered neither alive nor pulverized. But one or the other state would be brought about by observation.

If one rejects this view, then one has to assume that the function expresses no real fact but the essence of our knowledge with regard to a fact. This is Born's interpretation which is probably shared today by most theorists. But then the natural laws which can be formulated do not refer to the temporal changes of something that exists but to the temporal changes of the essence of our justified expectations.

Both viewpoints are without logical flaw; but I am unable to believe that one of these viewpoints will finally stand the test of time.

There is also the mystic (Bohr) who generally sees it as unscientific to ask for something existing independently of what is observed, i.e., the question whether the cat is alive or not, at a specific time before the observation. Then both views merge in a white fog, in which, I, too, do not feel any better than about one of the previously mentioned views which take issue on the concept of reality.[43]

The complete quantum theory Einstein found lacking would be one in which it would not be necessary to observe an individual case as a whole - which "one could not reasonably demand", as he added. He wrote all this in order to make the addressee of this letter understand the viewpoint that had "driven into deep solitude." (August 9, 1939).[44] Meanwhile he no longer harbored any illusion that he could convince Schrödinger of this viewpoint.

4. EINSTEIN'S WITHDRAWAL AND SCHRÖDINGER'S OFFENSIVE

Although new issues made the feelings of the leading physicists run high after the the war, in particular that of the responsibility of the scientist in light of the challenges of the nuclear age, the discussions on fundamental issues of non-relativistic quantum mechanics did not subside but were, on the contrary, rekindled. The Einstein-Bohr debate was continued by the representatives of a new generation of physicists.

In 1949, in a highly interesting letter to Arnold Sommerfeld, Erwin Schrödinger answered for himself the question "Einstein or Bohr?" as follows:

> Whereas I completely agree with Einstein and share the universal admiration for the amazingly extensive and intensive work of his lifetime which is truly exceptional (our regrettable quarrel is purely personal), I have the deepest amiable feelings for Niels Bohr and, of course, the highest respect for the work in which he has been leading force since 1912 and which will also continue to be fundamental for the more recent development which set in in 1925 with de Broglie and Heisenberg. At the same time, I find the influence erroneous and regrettable which he has exerted himself due to his tremendous authority on this more recent development, above all, by inventing some catchwords, such as complementarity, direct influence of the observer on what is to be observed, blurred boundary between subject and object, unfeasibility of models, science = recipe for prophecy etc. all the way to the physical theory of free will...
> No one seems to notice the above-mentioned complex of catchwords has persisted for already two decades with only slight modulations, and that it has not led to any single tangible success - contrary to Bohr's earlier intuitions (including the principle of correspondence). In this respect, I always remember Anderson's fairy tale about the 'king's new clothes'. Everybody believes that there must be something to it because everyone pretends as if they see something behind it. Quite seriously, I ask one to consider whether it is not merely words that have presented themselves at the right time, where concepts are lacking. Anyhow, it seems to be that during these twenty years of talking and also because of it, the most important finding of the 'new mechanics', the most important physical and philosophical clue has fallen into oblivion. Every physicist knows this but one does not speak about it, thinks it to be less important, and couches the matter in comfortable but concealing jargon. *The corpuscle is not an identifiable individual...*[45]

Einstein was aware of his intellectual kinship with Schrödinger, and wrote to him:

> You are the only contemporary physicist, besides Laue, who sees that one cannot get around the assumption of reality - if only one is honest (August, 22., 1950).[46]

As in his work published with Podolsky and Rosen in 1935, he held the view that quantum mechanics was not a complete theory and therefore in need of being extended:

> If one wants to regard quantum theory as final (in principle), then one has to believe that a more complete description would be pointless because no laws would exist for it. If this were so, then physics could only claim some interest from merchants and engineers; the whole matter would be a sad mess.[47]

Einstein agreed with Schrödinger on his criticism of the particle concept and stated:

> Of all our instruments only the field concept remains; but only the devil knows, whether it will endure. I think it is worthwhile preserving it, i.e., the continuum, as long as one has no real watertight reasons against it.[48]

Schrödinger was bound to see this as a confirmation of his intuitions regarding natural philosophy, speculations with which he had already set out to elaborate wave mechanics. In the meantime he was concerned with the physical program of an all-embracing understanding of matter in terms of waves. He knew:

the attempt to understand matter in terms of waves can succeed only if one understands all matter involved here in the same way. In this respect, I mean by 'in terms of waves' also 'continuous,' without any quantum leaps and bounds.[49]

Schrödinger continued to be hostile towards Bohr's complementarity argument, which he called a "clever way out"[50], without really exploring its more far-reaching ontological and epistemological implications. His criticism of Bohr's complementarity was a critique of a dualism which Bohr did not advocate in this matter at all.

For Schrödinger it was a fact that physics had to be changed as a consequence of the special methodological status it had acquired through Bohr,

It was a principle taken from everyday life and adopted to science. It has been held upright to date by all other sciences except for physics. Tacitly subscribing in a matter-of-fact way to that notion, it is certainly incorrect when we make up something to ourselves about a fact or a phenomenon, if this idea does not conform with what we can certainly learn about this thing, with whatever instruments and means of research. If this failed, then one went to great lengths over many years to obtain this conformity, and is still doing this - in other sciences. [51]

In his attempts to constructively transcend Bohr's viewpoint Schrödinger insisted less and less on new, more adequate concepts from the early 50s on. His articles published in 1952 and 1955, "Are there Quantum Jumps?" and "The Philosophy of Experiment" conveyed the impression that the notion of waves formulated in them had already provided the means for a unified physical picture of the world based on complementarity and dualism.

In this context he again attacked Born's probability interpretation of the wave function but this time his criticism was more scathing. A letter to Bohr written in 1952, for instance, reads:

A lecture course that I gave this winter on the current views in quantum mechanics has convinced me definitively that they are inadequate from the outset, viz. from Born's probability interpretation, which I disliked from the first moment on and have ever since. So I have decided to take a firm stand against it, pointing out its philosophical shortcomings. I have little hope of convincing many people now, the credo is too firmly established.[52]

But then Born found a completely unexpected strategy of defense, questioning whether the Laplacean determinism was really adequate to classical mechanics.

Meanwhile, modern theories of chaos furnish the convincing proof that the deterministic view corresponding to mechanical materialism is an inadmissible simplification of non-linear problems of motion lending themselves to classical treatment. Born's considerations already moved in this direction in the fifties.[53] At the time, he proudly wrote to Schrödinger:

It is now time for us to change from defense to the offensive. I shall not leave out any occasion to show that classical mechanics is just as useless in a deterministic manner as a porcupine or hedgehog is as toilet paper.[54]

Schrödinger's investigations from that time yielded no cognitive advance in physics and unnecessarily confused some of the already clarified problems. For this reason the "wave philosophy" endorsed by him[55] met with the protest of many physicists. Einstein, who had meanwhile commenced his retreat into a safe stalemate position in the debate with Bohr, restricting himself to the assertion that a better theory than the present quantum theory had to be found, distanced himself from Schrödinger's futile attempts to draft the outlines of this future theory.[56] Schrödinger and Einstein also had different reasons for rejecting Born's interpretation of the wave function.

For Einstein an ontology in which chance also plays a major role was utterly unacceptable at first. When it became more and more obvious to him that quantum mechanics no longer admits any classical ontology, his critique shifted its focus from the interpretation of the physical theory, which was very successful in practice, to the interrelation between ontology and epistemology in Bohr's complementarity argument. Having seen that his own attempts had failed, he did not provide any specification whatsoever as to how to imagine a return to a theory in the sense of classical physics.

As opposed to Einstein, Schrödinger was indeed willing to accept an ontology where chance plays a major role. But he was not ready to accept discontinuities as constituent features within his ontology (which was no problem for Einstein). Schrödinger felt that the success of wave mechanics confirmed his conviction with regard to continuity and, despite repeated setbacks, he would not desist from searching for a uniform ontology of nature complying with the principle of continuity. His critique of Born's interpretation of quantum mechanics was from the very outset directed against a renunciation of ontology and against the assumption of discontinuities in nature. But the latter point was also bound to remind Einstein of Schrödinger's original rejection of his light quanta hypothesis.

In 1955, W. Pauli pointed out the different viewpoints within the "opposition against quantum mechanics:"

Among the members of this opposition I understand *Einstein* best, at least I believe so. A year ago I saw him again in Princeton; I did *not* share his opinion but I found he could express well what *he* wants. Incidentally, he did *not agree at all* with the *other* "opposition members!"[57]

Pauli saw the fundamental reason for the differences in the evaluation of quantum mechanics in the different standpoints on "viewing the position of man (i.e., in physics: of the observer) in Nature." With regard to Schrödinger he stated:

Anyhow, the reason indicated ('position of Man in Nature'- for Einstein 'reality postulates') applies to the divergencies of opinion on quantum mechanics of *Einstein* and myself.[58]

Pauli's cautious allusion to the decisive point in the divergence from Schrödinger's view was justified insofar as his statement on the role of the observer created many riddles to be solved. In 1947, he had, for instance, written:

A fundamental feature of our picture of the world...is the exclusion or leaving out of the subject of cognition who falls back into the role of the detached observer...The implication of this simplification is that in a conception of the world formed in that way the viewer himself, his sensual qualities, his self-awareness, do not appear in adequate form, they are only imperfect spare parts represented as such, made of cardboard, as it were. Indeed in this world atoms, electrons and light quanta exist but there is nothing blue, green, bitter, sweet, no consciousness (for it stands outside), no will.[59]

And immediately after that a strange picture was introduced, implying that the representative of consciousness, i.e., the brain, is a late...ephemeral phenomenon in the evolutionary process of life on earth, a phenomenon which will vanish again a few years after the sun cools down: "But outside the viewer stands and sees the sun cool down."[60] In view of this incongruency, is not the old picture repeatedly used by Niels Bohr to be preferred, in which man is both the actor and the spectator on the stage of life?

Bohr responded in a very direct manner, although he had the "strange feeling" of having to teach quantum mechanics to a colleague who had quite substantially contributed to its development. As a response to "Are there Quantum Jumps?" he wrote:

Of course, we all know today that no precise time description that can be applied to a process which in a well-defined manner can be described as a transfer of energy, and that therefore neither the word 'jump' nor a reference to such a situation to which the old saying 'natura non facit saltus' belongs is adequate in such context. If I was right on any point in those days it was perhaps most in my reluctance to enter on a detailed analysis of the process referred to as complete transitions between stationary energy states...your presentation in the article and your attempts to reduce all analysis of experience to pseudo-classical resonance-effects do not give the right impression of neither the situation in atomic physics nor of the scope of the quantum mechanical methods.
What we in any special case will have to regard as the object under investigation and to which the variables in the wave equation refer, is determined by the experimental arrangement specified by the parameters entering in the wave equation. Just the great variety of possibilities in these respects seems to have given rise to much confusion, and I cannot see otherwise than the 'hope' expressed by Einstein as regards reverting to a casual description, and your 'belief' that it should even be possible to reach this goal in a comparatively simple manner both have their roots in insufficient attention to the conditions and possibilities of observations.[61]

In their opposition, Born, Heisenberg and Pauli were even more outspoken. Born wrote laconically:

For me it is farce to think that one should even describe my penholder, my own body, etc. exclusively in wave language, if also the employment of the particle language is graciously conceded as being a useful abbreviation. You do not believe this yourself...[62]

Heisenberg put forth a frequently expressed objection which had been rejected by Schrödinger time and again:

One can certainly discuss a great deal with the words one uses to interpret the usual interpretation of quantum theory. In this connection, one arrives at the boundary between philosophy and physics, and many physicists have not learnt how to move the right way. But these questions no longer have to do with the physical interpretation of quantum theory, and temporarily I am convinced that this theory contains no contradictions whatsoever.[63]

Pauli finally made fun of Schrödinger's "declaration of war against quantum mechanics":

I am convinced myself that a return to Newton's laws of nature in 1927 is not possible: *in this way* it is certainly *not* possible!
On the contrary, I would be pleased if somebody could invent something really new, in the fundamental concepts (even if not in its conclusions) different *both* from 'classical physics' and those which are valid at present.

I do not cling to the existing state of things or only *relatively* speaking to what has preceded it. I am rather progressive-minded and not regressive-minded - and consider the latter attitude to be psychically quite unhealthy, especially also for you![64]

Under this line there is a note added in unmistakable handwriting: "Moron! E.S."

Schrödinger's reactions to his critics went from amusement to despair. For instance, in 1953 he wrote in a letter to Born: "

For me it is particularly amusing, how your criticism attacks me once from the right side, and then from the left. On the one hand, I am a raving revolutionary, 'obviously a violation of historical continuity.' On the other hand, Pauli (in the private letter that you quoted) counts me among 'all reactionary efforts.' Can you blame me if the idea occurs to me, perhaps I am neither this nor that but simply sensible?[65]

He sounded somewhat different tones about eighteen months later in another letter to Pauli:

I know that most people, even some of my friends (e.g., Max Born) think I am simply so in love with my old waves from 1926 that I cannot get away from them. At least do you should not believe so no matter what I may have said that may sound like that.[66]

At the end of his life Schrödinger felt - like Einstein - that he was becoming increasingly isolated in the scientific world. Only rarely did he find physicists with whom he could agree, like with H. Margenau, on the fundamental discussions regarding quantum mechanics:

your views...are closer to mine than those of anybody else; indeed, you are perhaps the only person who feels certain fundamental difficulties with me, while the whole younger school of physicists bypasses them.[67]

Schrödinger emphasized that the wave function by necessity had two properties which can only be reconciled with great difficulties, notably the following:

...wave function sometimes as embodying a 'reality', sometimes only information, smearing over this distinction the lime of positivist philosophy ('how, prithee distinguish between reality and our knowledge about it?'). Naturally, information changes discontinuously when new information is gained. But how waves of information...should also form interference patterns, is not easy to say.[68]

Some of the most influential philosophers of science took a different stand from that of the physicists. B. Russell for example expressed his delight in a letter to Schrödinger:

I am very much comforted by your being 'unable to understand contemporary theories in physics'. It makes me hope that perhaps they are as misguided as they seem.[69]

The father of wave mechanics corresponded with K. R. Popper between 1952 and 1953, seeking to improve the argumentation of the philosopher against the Copenhagen interpretation of quantum mechanics.[70] However, he did not succeed in doing so.

NOTES

1 I would like to thank all legal successors for allowing me to quote so extensively from the material cited here. The director of the Niel Bohr Archives in Copenhagem, Mr. Erik Rüdinger, was kind enough to give me access to the "Bohr Scientific Correspondence" (BSC) and the "Archives for the History of Quantum Physics" (AHQP). In the quotation reference, the numbers before the comma refer to a specific microfilm, whereas the number after the comma refers to a specific section of the microfilm (e.g., AHQP 37,5 means: "Archives for the History of Quantum Physics" microfilm number 37, section 5).

2 N. Bohr, Discussion with Einstein on epistemological problems in atomic physics, in: P. A. Schilpp (ed.), *Albert Einstein: Philosopher and Scientist*, Evaston, Illinois 1949, p. 199; A. Einstein, remarks on the works included in this volume, ibid., p. 494.

3 For detailed comments of the author on this issue as well as further literature see: U. Röseberg, *Szenarium einer Revolution.* Nichtrelativistische Quantenmechanik und philosophische Widerspruchsproblematik, Berlin 1984.

4 E. Schrödinger, *Mein Leben, meine Weltansicht*, Vienna, 1985, p. 71.

5 op.cit., p. 73.

6 M. Born, in a letter to E. Schrödinger, 2 17.1951, in: AHQP 37,3.

7 E. Schrödinger, in a letter to Franz Theodor Csokor, 10.17.1960, in: E. Schrödinger, *Mein Leben, meine Weltansicht*, a.a.O., p. 2.

8 ibid., p. 55.

9 N. Bohr, in a letter to G. Hevesy, 1.25.1926, in: BSC 11,3.

10 W. Pauli, in a letter to A. Sommerfeld, 2.9.1926, in: *W. Pauli, Scientific Correspondence with Bohr, Einstein, Heisenberg, et al.* (referred to in the following as PWB), vol.1, New York, Heidelberg, Berlin 1979, p. 293.

11 W. Pauli, in a letter to E. Schrödinger, 5.24.1926, in PWB 1, p. 324-326.

12 M. Planck, in a letter to E. Schrödinger, 4.2.1926, in: *Schrödinger, Planck, Einstein, Lorentz, Letters on Wave Mechanics*, Vienna 1963, p. 3.

13 A. Einstein, in a letter to E. Schrödinger, 4.26.1926, in: Schrödinger et. al., *Letters on Wave Mechanics*, loc.cit. p. 26.

14 cf. W. Heisenberg, *Der Teil und das Ganze*, Munich 1969, p. 104 ff.

15 H. A. Kramers in a letter to E. Schrödinger, 6.15.1926, in: AHQP 41,8.

16 E. Schrödinger, in a letter to M. Born, 11.2.1926, in: AHQP 41,7.

17 E. Schrödinger, in a letter, 10.23.1926, in: N. Bohr, *Collected Works* (referred to the in following as NBCW), vol. 6, Amsterdam, New York, Oxford, Tokyo 1985, p. 459-461.

18 Many writers on the subject have pointed out that Schrödinger, in 1924, expressly welcomed the new radiation theory of Bohr, Kramers and Slater referring to Exner's position on the universality of statistical laws which was a continuation of Boltzmann's ideas. At the time he wrote to Bohr that he had long been considering the idea "that the basis of our statistics is probably not microscopic 'regularity,' but rather perhaps 'absolute chance' and that perhaps even energy and impulse theorems are only valid in a statistical sense." (E. Schrödinger, in a letter to N. Bohr, 5.24.1924, in: NBCW 5, Amsterdam... 1984, p. 490.) After the strict validity of the energy and impulse law at least for the area being dealt with here, had been experimentally proven, Schrödinger no longer endorsed the general validity of statistical laws with his initial vehemence but at the same time he did not claim the opposite.

19 E. Schrödinger, in a letter to W.Wien, 10.21.1926, in: AHQP 41,11.

20 cf. M. Born, in a letter to E. Schrödinger, 5.16.1927, in: AHQP 41,7.

21 ibid.

22 M. Born, in a letter to E. Schrödinger, 11.6.1926, in: AHQP 41,7.

23 E. Schrödinger, in a letter to G. Joos, 11.6.1926, in: AHQP 41,7.

24 ibid.

25 ibid.

26 E. Schrödinger, in a letter to N. Bohr, 11.25.1926, in: BSC 16,2.

27 E. Schrödinger, in a letter to H. A. Kramers, 11.19.1926, in: AHQP 41,8.
28 E. Schrödinger, in a letter to M. Born, 6.8.1927, in: AHQP 41,7.
29 E. Schrödinger, in a letter to M. Born, 5.6.1927, in: AHQP 41,7.
30 N. Bohr, in a letter to A. Einstein, 4.13.1927, in: NBCW 6, p. 419.
31 cf. U. Röseberg, *Niels Bohr, Leben und Werk eines Atomphysikers*, Berlin 1985, pp. 156 ff.
32 E. Schrödinger, in a letter to N. Bohr, 5.5.1928, NBCW 6, p. 464.
33 N. Bohr, in a letter to E. Schrödinger, 5.23.1928, NBCW 6, p. 465.
34 A. Einstein, in a letter to E. Schrödinger, in: *Schrödinger, Planck, Einstein, Lorentz, Letters on Wave Mechanics*, New York 1967.
35 E. Schrödinger, in a letter to N. Bohr, 9.25.1930, in: BSC 25,3.
36 E. Schrödinger, in a letter to N. Bohr, 4.29.1931, in: BSC 25,3.
37 ibid.
38 N. Bohr, in a letter to E. Schrödinger, 5.8.1931, in: BSC 25,3.
39 ibid.
40 W. Heisenberg, in a letter to N. Bohr, 9.29.1935, in BSC 20.2.
41 N. Bohr, in a letter to Ph. Frank, 1.14.1936, in: BSC 19,3.
42 A. Einstein, in a letter to E. Schrödinger, 3.23.1936, in: AHQP 37,5. This letter is quoted with friendly permission of the Albert Einstein Archive (Hebrew University of Jerusalem.)
43 A. Einstein, in a letter to E. Schrödinger, 8.9.1939, in: *Schrödinger, Planck, Einstein, Lorentz, Letters on Wave Mechanics*, pp. 32-33.
44 ibid.
45 E. Schrödinger, in a letter to A. Sommerfeld, 2.13.1949, in: AHQP 37,12.
46 A. Einstein, in a letter, 12.22.1950, in: *Schrödinger, Planck, Einstein, Lorentz*, op.cit.
47 ibid.
48 ibid., pp. 36-37.
49 E. Schrödinger, in a letter to M. v. Laue, 3.11.1951, in: AHQP 37,8.
50 E. Schrödinger, in a letter to M. Born, 2.11.1951, in: AHQP 37,3.
51 ibid.
52 E. Schrödinger, in a letter to N. Bohr, 6.3.1952, in: BSC 32,3.
53 cf. U. Röseberg, *Determinismus und Physik*, Berlin 1975, p. 76 ff.
54 M. Born, in a letter to E. Schrödinger, 1.11.1954, in: AHQP 37,3.
55 M. Born, in a letter to E. Schrödinger, 7.25.1952, in: AHQP 37.
56 cf. *A. Einstein, M. Born, Correspondence 1916 - 1955*, Hamburg 1972, p. 200.
57 W. Pauli, in a letter to E. Schrödinger, in: AHQP 37,10.
58 ibid.
59 E. Schrödinger, in a letter to A. March, in: AHQP 37,9.
60 ibid.
61 N. Bohr, in an undated letter (1952?) to E. Schrödinger, in: BSC 32,3.
62 M. Born, in a letter to E. Schrödinger, 7.25.1952, in: AHQP 37,3.
63 W. Heisenberg, in a letter to E. Schrödinger, 10.30.1952, in: AHQP 37,7.
64 W. Pauli, in a letter to Schrödinger, 6.26.1952, in: AHQP 37,10.
65 E. Schrödinger, in a letter to M. Born, 4.11.1953, in: AHQO 37,3.
66 E. Schrödinger, in a letter to W. Pauli, 1.31.1955, in: AHQP 37,10.
67 E. Schrödinger, in a letter to H. Margenau, 4.12.1955, in: AHQP 37,9.
68 ibid.
69 B. Russell, in a letter to E. Schrödinger, 5.23.1952, in: AHQP 37,11.
70 AHQP 37,10 includes 6 letters with a total of 20 pages of correspondence between Popper and Schrödinger.

THE NOTION OF CONSCIOUSNESS IN SCHRÖDINGER'S PHILOSOPHY OF NATURE

HANS POSER

Institut für Philosophie, Wissenschaftstheorie, Wissenschafts- und Technikgeschichte der Technischen Universität Berlin Ernst-Reuter-Platz 7, 1000 Berlin 10, Germany

> "If metaphysics was to really be abolished, arts and sciences as well would be reduced to lifeless siliceous skeletons, incapable of the slightest further development."
>
> E. Schrödinger: *Meine Weltansicht*, p. 14

> "A purely rational world view is an absurdity."
>
> E. Schrödinger: *Mein Leben*, p. 2

1. INTRODUCTION

Even in light of Thomas S. Kuhn's critique, it is still generally assumed today that the historical development of science is guided by paradigms, which first and foremost define its issues and methods. Even if this view is not necessarily closer to the truth than others, it has become a paradigm of the diachronic understanding of science, under which the respective paradigm of a discipline is subsumed. But whereas Kuhn and later on I. Lakatos, L. Laudan, and even P. Feyerabend put emphasis on the limitations of the preliminary decisions constituting a science, one must be aware of the fact that these are embedded in a wide network of values and fundamental views of a time, which, on the one hand, is the presupposition for changes within a specific science on the basis of argumentation, whereas, on the other hand, this network is influenced by new scientific findings. This general framework, i.e., the *prescientific world view*, changes much slower than paradigms. This can be seen in the teleological view on which each explanation and interpretation of nature, from Aristotles to the Middle Ages has been based, and later in the causal view, expressed in the strictest sense by the Laplacean Demon, a view, which the evolutionistic view seems to have suppressed today.

A reflection on the framework of a world view leads us into the realm of *natural philosophy*. In the past, the term natural philosophy has had two meanings: on the one hand, empirical knowledge of nature (think of the Newtonian title "Principia mathematica philosophiae naturalis") and on the other, the attempt to clarify the metaphysical presuppositions underlying such kwowledge (as formulated, e.g., in F. W. Schelling's "Introduction to a Concept of a System of Natural Philosophy"). In this context we are dealing with natural philosophy in its second

J. Götschl (ed.), Erwin Schrödinger's World View, 153–168.

meaning. As a consequence of the critique of metaphysics (assuming a metaphysicist to be someone who, in a dark room, is searching a non-existing cat), natural philosophy gradually became philosophy of science in that it is limited to methodology and to Kuhnian paradigms. Kuhn's view of paradigm was the historian's answer to Carnapian verifiability and Popperian falsificationism. But in concentrating only on the empirical objections to a theory, Kuhn believes that a new and an old paradigm are incommensurable, whereas conceptual discussions (e.g., as between Leibniz and Newton concerning the structure of space and time and the possibility or impossibility of gravitation as action at a distance), are not taken into account, let alone hermeneutic problems. Meanwhile it has become clear that this limitation of natural philosophy is too narrow. Yet understood again as a metaphysics of nature, it is opposed by those old and well-founded arguments against the ability of metaphysics to be able to give absolute answers. All of those allegedly absolute, final and a priori foundations of science turned out to be untenable - be it Descartes' deduction of the laws of motion from attributes of God, Spinoza's intention to clarify principles of extension on a purely conceptual level, the Leibnizian proof of at least some of the laws of dynamics by the means of the principle of the best, Kant's transcendental underpinnings of Newtonian physics or Schelling's foundation of his speculative physics on a concept of becoming. But even if such foundations are rejected, the extrapolation of empirical theories to world views took place throughout the positivistic 19th century from Oerstedt and Ritter to Mach and von Helmholtz. The radical changes in the principles of physics due to the theory of relativity and quantum theory were accompanied by reflections leading beyond empirical foundations. This can be said to be true of Einstein as well as of Heisenberg, P. Jordan and C. Fr. von Weizsäcker: scientists themselves contributed to natural philosophy, thereby influencing the precision and the transformation of the prescientific world view preceding the paradigms and hypotheses of empirical science.

Schrödinger's reflections follow this line as well, but they also differ from it to the extent that he, who for some time considered giving up his professorship of theoretical physics for philosophy, did not aim at extrapolation, but at a universal conception embracing the individual subject and the world as a sum of various relationships. This conception is founded on the same global claim as Carl Friedrich von Weizsäcker's *Aufbau der Physik* with its transcendental thrust. Moreover, Schrödinger, the founder of wave mechanics, also placed great emphasis on a quantum theoretical understanding of life and of biological evolution. He is among those who introduced a new view which has been developed today in the work of R. Riedl, M. Eigen, and I. Prigogine - thinkers who do not so attempt to establish universal laws as to understand our unique universe, this system of planets with its singular species, that is to say, to restore the position of man. Schrödinger, however, tries to go even further: his interest is also the subject, the ego, the self, the 'I', as Schrödinger says. Natural philosophy should thus not only take man into account but also the 'I' as its point of departure. It should deal with the *questions* posed by Descartes and Husserl instead of immediately postulating the res cognitans or the transcendental I, or even excluding man, as traditional natural philosophy. This means that we should try to reconstruct what Schrödinger himself only sketched: Schrödinger's philosophy of nature and his understanding of consciousness.

2. METAPHYSICS AND EXPERIENCE

In Schrödinger's opinion, Kant did away with the possibility of theoretical metaphysics as a priori foundation of knowledge. At the same time he claims that "metaphysics is the indispensable basis of our universal as well as our individual knowledge".[1] For even if metaphysics is no longer part of the building of science, it is the scaffolding needed to erect a building and to continue construction.[2] Underlying each attempt to probe further, to attain transcendence by way of reasoning and to formulate new findings in words, metaphysics is the invariable presupposition of all knowledge and science. Thus, the contrasting and contradicting positions of different metaphysical approaches are not the "babble" of "simpletons", but could also be a well-founded view reflecting various perspectives.[3]

Against this background Schrödinger develops what he calls "the key issue", which does not let the mind rest until it is answered, for we cannot answer the following questions with a simple "yes" or "no": "1. Does an 'I' exist? 2. Does the world exist *along with me*? 3. Does the 'I' end with the death of the body? 4. Does the world come to an end with the death of my body?"[4]

These questions all focus on the relation of the 'I' and the world, or on the question "What am I?" This sounds like Kant's question "What is the human being?", which he sees as the ultimate question of philosophy. Schrödinger, however, has something different in mind: he wants to use this 'I' of mine as a point of departure for understanding the 'I'. His approach bears resemblence to a Cartesian meditation.

Schrödinger clearly recognizes that these questions cannot be answered through discursive thinking alone, for it is impossible to grasp the last foundation by means of argumentation and language. The only other option perhaps is to use pictures and metaphors, i.e., to pursue argumentative thinking as far as possible. And from there one can continue with metaphoric thinking, which ultimately leads to "a highly desirable completion of our world view", as Schrödinger puts it.[5] Here he goes beyond the traditional demand for the absoluteness of physics, in the same way Leibniz did in calling his monadology a hypothesis (i.e., one which allows the view of the world as an empire of causes to coexist with the view of the world as an empire of aims of God), or as Whitehead with his view of metaphysics as a cosmological scheme of thought open to criticism and alterations.

Schrödinger never gave up the problem of the relation between the 'I' and the world. He discusses it in the fall of 1925 and continues to discuss it up until his Alpbach article "What is real?" (1960), which he included in his *My World View*. Since he uses the same formulations for several decades in other articles as well, we can make use of these ideas without referring to the date.

Schrödinger never intended for his philosophical ideas to constitute a metaphysical system. The following, therefore, must be understood as an attempt to reconstruct his position and his arguments.

The answer to the question "What am I?" - and this is Schrödinger's ceterum censeo - cannot be given by science, no matter how far it ever progresses, for behind it there is also the other question as to the meaning of life. He thus does not deliberately focus his discussion on problems such as "acausality, wave mechanics, indeterminacy, models of global expansion and creatio continua, etc"[6] which one would expect in a metaphysics of nature. It is clear that he does not want to see philosophy as a direct continuation of quantum mechanics with different means, and he attacks those who try to do so.[7] On the other hand, we have seen that for him metaphysics is the scaffolding of the building of science. It thus comes as no surprise that Schrödinger seeks to re-establish the links between metaphysics and the empirical sciences, in particular physics, but also biology and physiology, in his global perspective.

This frame of reference can be applied to our present view of the empirical sciences as well as to that of philosophy: his philosophical reflections do not aim at establishing absolute foundations but at creating a coherent structure open to critical arguments (which is thus a part of our rational culture of scientific thinking). Such a structure differs fundamentally from empirical sciences in that it develops a world view which serves as orientation and as attribution of meaning and values. The framework is, as Ayer puts it, a meta-scientific system with a structural function: it enables us to understand science (understood as a primary system) as parts of a secondary order. Even if Schrödinger does not discuss in general terms how one can get from the one level to the other, we are able to envision this when we observe how he puts his ideas to work. This will be our next step.

3. DOES THE 'I' EXIST?

What do I mean when I say that "I am" ? I see myself as "that consciously thinking mental being which feels and identifies itself as 'I', namely as a person", Schrödinger writes.[8] This would be an I just in the broader sense of the Cartesian notion of cogitatio, referring to a thinking, wilful and perceptive I, in short, a spiritual I. But it is no isolated res cogitans, since, given the "linkage of all sensations of this 'I' with the material modifications of its own body"[9], I am forced to accept the existence of the body together with the 'I'. From the very outset one can exclude the possibility that this 'I' (as a soul) lives in the body as if the latter were a house (this idea of a ghost in the machine was parodied by G. Ryle). Consequently, one must accept that the 'I' and the world are "composed of the same empirical elements",[10] which means that the world does not exist beneath the I, but that *my own body is a complete part of the world.* Indeed it was Schrödinger who showed how our own body can be understood as a "pure mechanism that acts in accordance with the laws of nature", if the concept of entropy is extended to include "negative entropy" (or negentropy, as we say today).[11] This is the *physical perspective*; it is based on the well-known scientific mode of objectification, which lets us see ourselves from the outside as independent observers. However, the 'I' is more than just an objective something. My 'I' is also a being with sensations, volition and responsibility: the physical perspective, therefore, is too narrow.

If we replace the physical perspective by a *physiological perspective*, it becomes clear that when the body is destroyed the 'I' is destroyed as well. Seen from this perspective, the third of Schrödinger's first questions would be answered affirmatively. To accept this has several consequences. As we have seen, the physical view hinges on an external standpoint, which is not only inadequate with respect to my 'I', but also untenable. But if one asks from the perspective of the 'I' what the world is, one is forced to admit that it can only be grasped as a sensation. From this it follows that *the world (as a sensation) is a complex part of my I*! However, it would be "completely ridiculous" to assume that my death is the end of the world![12] To avoid this consquence we have to find the implicit presuppositions of both views to decide which has to be given up and which one altered in order to avoid this consequence, i.e., to look for further ways to analyse the 'I'.

As both the physical and the physiological views show, this body as my unique body, defined by the totality of its properties, cannot be distinguished from all the others so that a soul in the sense of an 'I' can be attributed to it. Rather, the distinction in question must be made the other way round: The separation of body and its properties presupposes the uniqueness of the 'I' as a mental 'I'. Therefore we should try to find a further perspective and start from the phenomena of the soul. When I see a tree, the substratum of my sensation is neither what is going on in my central nervous system, nor the sensation of a tree as a tree, nor the tree as such. The radical solution of E. Mach, R. Avenarius, and W. Schuppe, which was a continuation of the tradition of Berkeley's esse est percipi, was to only take the *perception* as a given fact. This implies that the tree and the perception of the tree are the same objects.[13] In principle Schrödinger goes along with this tradition, since his solution allows one and the same element (which Schrödinger sometimes calls "elements of consciousness"[14]), to be seen as both an object of the external world and a constituent of the I in form of perceptions. That what really exists, are just these elements of consciousness. To imagine the existence of an 'I' entails a false substantiation, since the 'I' is neither a substance nor a substratum; it only reflects a particular perspective which is by necessity my own.

The position which we have developed took the perceiving 'I' as its point of departure. In this respect it runs the risk of reverting to solipsism. But all the arguments pro and contra solipsism presuppose the familiar objective, external view of science. From this it immediately becomes clear that all these arguments do not take into account that this I, which is my point of departure, is *my* 'I'. The dilemma of solipsism results from the substantiation of my 'I', in that I erroneously see myself as an object.

Let us analyse the difficulties resulting from the following two different views: (a) If someone else perceives a tree as well, in which sense is it then the same tree? And (2): How can one and the same tree as a perception (not a an object) be a constituent of different "consciousnesses"? Indeed the positivism of Avenarius and Schuppe accepts a *unity of consciousness*. This, however, contradicts the intuition that our thinking of an 'I' is independent of thinking of another 'I'. But to Schrödinger there is no problem here at all. We are all familiar with the intuition that two people are thinking the same thing - for instance, when they are solving the same arithmetical equation. Why then should two people not have the same perception of a tree? But the situation is a paradoxical one when I recognize that this seeming symmetry - two people with the same or with different thoughts - reflects an outside view as well, whereas correctly seen, one of them is always 'I'! This leads to inextricable asymmetry and perspectivity.[15] The "essential philosophical difficulty here is", therefore, "the variety of perceiving and thinking individuals."[16] Here science can no longer guide us, since the external view is its methodological principle. This also explains why Descartes and Husserl lead in the opposite direction, namely from the cogito to the sciences. This wc must kccp in mind in our analysis of the 'I'.

As a first and preliminary conclusion we can say that the 'I' in Schrödinger's view is a perceiving and thinking 'I'. In differentiating the second of Schrödinger's introductory questions, we get three classical problems which have to be solved step by step: a) How does the 'I' relate to the others? b) How does the 'I' (as consciousness) relate to the external world? c) How are body and soul (or consciousness) interrelated? These questions will be explored in the following sections.

4. THE UNITY OF THE 'I'S

What makes one 'I' different from the others? Schrödinger illustrates his approach with the way the Alps are experienced. This experience leads me to the question why 'I', who has this experience should be different from someone else who has seen and experienced this landscape from the same point of view earlier. Isn't it so that this feeling is identical for everyone and as persistant as the Alps? (Wilhelm Dilthey would have rejected this example. However, in order to explain how two different persons can understand something in the same way, hermeneutics must make a claim equivalent to Schrödinger's!) If this is the case, then, Schrödinger concludes, "this experience, this feeling and this will is eternal and unchangeable in its essence". Moreover: "You - as well as each conscious being for himself - you are at every time everything (Du bist alles in allem)"![17] Thus my life is not a part of the world, but it is the world in toto as an everlasting present now: "All consciousness - Schrödinger declares, wanting to be understood literally - is essentially *one*!"[18] This is the main thesis of Schrödinger's metaphysics. With such a philosophical monism he wants to facilitate that "most desired completion", which, for him, would solve the contradictions encountered in trying to understand the 'I'.

Monistic positions are far from evident. In the history of thinking they have been put forward to provide completely new and divergent world views in order to overcome seemingly irresolvable contradictions. This tradition shows two aspects which Schrödinger combines in his search for a unity. One, the animation of the universe: Schrödinger cites the hylozoism of the Ionian philosophers of nature, where mind and matter form an inseparable unity, Spinoza and his unity of cogitatio and extensio and G. Th. Fechner's panpsychism.[19] The second aspect is the unity of consciousness which Schrödinger sees maintained in the Islamic tradition - he quotes Aziz Nasifi [20] - and in Spinoza's pantheism. But the shortcoming of these positions, so clearly visible in Spinoza, is, according to Schrödinger, the fact that I am always forced to understand myself as a limited part of the whole (in Spinoza as a finite modification of the infinite), whereas the question remains unanswered as to how insight into the part-whole relation is possible and how I can distinguish myself objectively as a part from that whole.[21] To illustrate his answer Schrödinger uses a metaphor image of reality taken from the philosophy of Vedanta, in which unique reality is thought to be represented by the many hundred facets of a crystal. For him, this metaphor expresses the Brahmanian formula "tat twam asi (that is you)", which he translates freely as "I am this whole universe."[22] Along with this the interrelation between individuals is expressed, for on the one hand they have a unique root, whereas on the other, there are different facets of one and the same thing. Since this idea only exists as a metaphor, we have to develop its analytical content and study the way it can be applied to describe reality.

Before doing so, a general point of contention should be discussed: How can one seriously see this view as a continuation of the scepticism which had been the guiding force of the empiristic as well as of the scientific tradition? Let us briefly forget Schrödinger and remember that a similar view is held today by very different thinkers, as for instance, C. Fr. von Weizsäcker and Fr. Capra, even if their approaches differ. They both believe that only by striving for unification can we arrive at a new understanding of the relation between man and world. Capra, for instance, postulates a complementation of scientific rationality, whereas von Weizsäcker interprets quantum theory in such general terms so that it includes the conditions of the possibility of knowledge laid down in consciousness: for if each process in the world is described by quantum mechanics as a development of the state vector which corresponds to Schrödinger's wave function, then this is a description of human knowledge as well, i.e., of the contents of consciousness. Quantum theory must then even include states of consciousness. [23] It is not mysticism or Ying and Yang but an interpretation of Schrödinger's physical findings that led von Weizsäcker to the thesis of the unity of consciousness, which Schrödinger postulates explicitly. Our discussion, therefore, does not just involve abstruse speculations but corroborates a central point in today's discussion of world view.

Let us return to Schrödinger. His conception is constructed from two sides like a bridge. On the one hand, he suggests evaluating metaphysical hypotheses on the basis of their potential to be the "most desirable completion" of thinking, referred to above. On the other hand, this "completion" is to be obtained through a purely holistic conception. Our question is, therefore, whether both sides of the bridge meet. Schrödinger cites four grounds, which, like his remarks on metaphysics, need not be seen as final justifications. Rather, they only suggest why it is plausible and coherent to think that all consciousness is essentially one.

1. By means of natural language we cannot form a plural of the notion of "consciousness" - and if we were to do so we would not know what it means. This is an important suggestion.[24] But we must remember that Schrödinger warns us several times not to interpret the results of language analysis as if they automatically correspond to the objects.

2. Physiology shows that the perception of different organs is manipulated by the brain at different locations, as if there were different "consciousnesses" at work - but we ultimately get a unity which has no organic counterpart. Exactly this can be seen as a model for the different elements of the consciousness expressed in the plurality of individuals.[25]

3. The separation and isolation of individuals within the line of ancestors is familiar to us; but why do I see myself as being the same before and after sleeping? Arguments for this identity point to our memories. Thus we see our instincts as functioning as a sort of "supraindividual memory" within the line of ancestors:[26] when we consider how they influence our actions, then one adopts a totally different point of view, assuming there is a unity of consciousness documented in a line of ancestors. In this line patterns of behavior are genetically transmitted which guarantee a unity beyond the single individual - just as the memory bridges sleep.

4. Generally, the interpretation of meaning, objectives, ethical and esthetic values and norms do not have a physical or physiological grouding. "*Nature* does not tribute respect to life," Schrödinger writes, alluding to A. Schweitzer:[27] Norms, values, and responsibility belong to the realm of the mind. And if all consciousness is essentially one, then this has direct consequences for ethics, for this unity allows us to characterize actions as morally good, independent of the individual.[28] For Schrödinger, this is the most important point, for it allows meaning to be attributed to life, something which is not possible only on the basis of empirical sciences.[29]

Schrödinger adds some further remarks to these points. "My conscious life", including my conviction of my uniqueness as an individual, "depends on a characteristic *constitution* and *function* of my soma and especially of my central nervous system", he notes. But both parts are not able to support this conviction, for the *functions* are genetically determined and thus shared by all who have a conscious mental life. Each body is a copy of it, serving as a "blueprint, masterbuilder and building material" for the next copy: "Where shall we define here the beginning of a new consciousness?"[30] With respect to function, the line of ancestors must, therefore, be extended to the whole of mankind!

We come to the same conclusion if we view *constitution* as the basis of individuality, for it, too, is not sufficient for individuation: I assume that another individual (with the same functional predispositions in a situation identical to my own), is similar to me, if it had undergone the same development that I did. This development is governed (1) by physical influences from the outside, and (2) by so-called mental influences, which are seen as determining factors of my central nervous system, are all tantamount to the influence of our ancestors. Patterns of behavior, which are passed down by cultural traditions, must be included along with those which are genetically transmitted. Schrödinger sums it up this way: "No I stands alone"[31] for, as a constituent of the line of ancestors, it is this very line! Therefore the death of an individual "does not put an end to anything essential about a life", nor does life begin with birth; birth is only like a slow awakening from a deep sleep.[32] Therefore, the second and third of Schrödinger's questions quoted at the beginning are also misleading!

To show that the separation of an I as individual has no natural basis, Schrödinger points to simple organisms like polyps which after dividing into two parts, reorganize themselves as two complete organisms - similar to the way we reproduce a missing link in thinking (e.g., when adding 5 + 7). Which kind of identity of consciousness is attributed to both parts after the division? Consciousness is not divisible and does not exist as a multiplicity; only certain perspectives, e.g., the acceptance of a plurality of individuals, make us - erroneously- believe we are thinking of consciousness in the plural sense. But this inference is an illusion, for the body of each single individual is some sort of "republic of cells and organs" parts of which show a high degree of independence - something which can also be observed in symbiotic communities among animals. Schrödinger would have seen the latest findings in genetic biology, e.g., the fact that the whole ground plan of an organism can be found in each cell, or the possibility of cloning, just as he did in his time with respect to organ transplantation. Biology thus teaches us "what we really are: a republic of cells with only very limited homogeneity, limitability and indivisability."[33]

Schrödinger does not underestimate the problems which immediately surface: Why do I have consciousness, but not the individual parts of me, why does my consciousness not "consist of the individual of the 'I's of the cells of my brain?"[34] Why do the single human 'I's not constitute a higher 'I' as a higher order republic? But these questions are misleading as well: They stem from the attempt to base 'I'-identity on the unity of the body, but vanish once one radically advocates the unity of consciousness. Consciousness has no parts, no 'I's, no clusters of 'I's - on the contrary. Each member is always only consciousness.

This metaphysical hypothesis is grounded on the one distinct experience that we "never and nowhere" actually find a manifoldness of consciousness.[35] What seems to be manifoldness of discrete entities is "nothing but an illusionary variety evolving from forms of appearance of this unique 'I'"[36], an illusion which is - in a Whiteheadian way of speaking - the consequence of a hasty substantiation.

5. THE EXTERNAL WORLD AND CONSCIOUSNESS

In the preceding sections the seeming manifoldness of 'I's has been seen as resulting in a unity of consciousness. Schrödinger, however, pursues this issue further. Even "the external world and consciousness are one and the same", for "the world is a construct of our perceptions, sensations, memories."[37] This far-reaching thesis is, in Schrödinger's view, the outcome of epistemological reflection on physics: If we question the objectivity of physics we find the necessary renunciation of an objective description of nature, for "our signs and formulas and the pictures associated with them do not describe an object independent from the observer, but only the relation subject:object" - a relation which "is, strictly speaking, the only genuine reality" we know.[38] But relations are - as the Scholastics already knew - only in mente. For Schrödinger, therefore, this view implies a refutation of the "hypothesis of the external world".[39] Schrödinger quotes Eddington's well-known story of the two desks (one familiar to us from everyday life and the other which has no sensory qualities, but primary holes between atoms) for which he sees an analogy. Following Sherrington he ironically states that if the external world exists, consciousness walks around in it as if it were a ghost. He concludes that one inevitably seems condemned to be a ghost, either in the objective external world of the scientist or consciousness itself, which in thinking builds the first one whereby it retreats from it.[40]

The "hypothesis of the reality of the external world" - the logical consequence of "objectivation" since the Greeks - is only a "simplification" to help us "master the problems of nature" by excluding the "subject of knowledge" from nature.[41] Precisely the position which made the classical sciences possible leads us to irresolvable antimonies.[42] Therefore, the separation of subject and object must be done away with, for "my mind and the world are formed out of the same substance."[43]

One may be inclined to understand Schrödinger's thesis as a by-product of modern physics. The theory of relativity resulted from a reflection on what is observable, and the quantum theory included the influence of the observer to the observed; this seems to destroy the classical distinction between subject and object. Capra has, indeed, based his thesis of the unity of consciousness on this. Schrödinger, however, always rejected Bohr's and Heisenberg's proposal that quantum theory implies the dissolution of the subject-object opposition. (The Greek atomists had already concluded that there is an interaction between both!) Instead, Schrödinger emphasizes that the two physical systems, observing subject and observed object, should not be confused with the epistemological difference between subject and object. "For the perceiving consciousness is no physical system at all and cannot enter into physical interaction with it." Here, in the domain of philosophy, "a reservatation of the term 'subject' for the perceiving consciousness could well be adequate."[44]

This does not imply that Schrödinger's reflections were totally isolated from his views of physics: the best example of a connection is his paradox known as Schrödinger's cat: a cat sitting in a closed box, which will be poisoned if, and only if, a single decaying atom of radium switches on a corresponding machine. If, after the half-life of radium, we take a look - without opening the box - at what has happened to our cat, then the wave function does not even allow us to say that the cat is either dead or alive - only that it is half dead and half alive at the same time. Therefore, Schrödinger believes that it makes no sense in the light of the holistic conception of wave mechanics to speak of isolated objects. It is perhaps more adequate to see Schrödinger's ontological standpoint in physics as the outcome of his philosophical reflections, for he arrived at his concept of the object in philosophy much earlier than in physics. His concept allows us to break with tradition and to comprehend the subject-object relation in a completely new way. This, then, would illustrate how philosophy can function as the scaffolding of physics.

As in his idea of the unity of consciousness, Schrödinger uses biological facts to bolster his thesis of the unity of the mind and the world, arguing that higher forms of mental life are "linked with the function of highly developed brains."[45] Here we come across a controversial question - emphatically discussed in theories of artificial intelligence: "which material processes are directly combined with consciousness".[46] For a solution in keeping with his conception of unity

Schrödinger starts out from empirical hypotheses, e.g., that the functions of the brain are most efficient for adaption to changing conditions of the environment "by natural selection or in another way" in order to optimize the chance of survival.[47] But this empirical hypothesis contradicts Schrödinger's view, i.e., that "it is consciousness through which the world is originally manifested, moreover, through which it exists".[48] How, then, can it evolve in an unconscious, e.g., non-existing world? Or as Schrödinger formulated in a different way: why should evolution lead higher developed animals to see themselves and the world "light up in the light of consciousness", or why should it lead to animals with consciousness at all?[49] The answer from a Christian point of view has always been that the highest step of the creation must necessarily include a conscious being which is capable of understanding the world as a mirror of God's wisdom, power and goodness. Once this was no longer seen as a sufficient metaphysical explanation the matter became more and more complex. Here Schrödinger follows Spinoza's line: "For each material process there is a correspondence like the correspondence between the vital process of the body and our consciousness".[50] One finds the same answer in Leibnizian monodology which extends the realm of perceptions, by including unconscious perceptions so that consciousness is attributed to the active part of the monad, whereas the passive one appears merely as a phenomenon. Consciousness, according to Schrödinger, emerges when an organism must react to a new situation. This applies to phylogenesis: "Only the individual characteristics of a single ontogenesis become conscious."[51]

Clearly, this is an extension of the traditional concept of consciousness, an extrapolation, which C. Fr. von Weizsäcker advocates in a similar way. It might be helpful to understand this new concept in the sense of the Leibnizian concept of active perceptions, as Schrödinger explains: We know that "each process in which we are consciously and actively involved sinks in and finally disappears from the sphere of consciousness, when it is repeated in the same way several times."[52] Schrödinger concludes: "Consciousness is combined with the *learning* of an organic substance; the organic *abilities* are not conscious". Put in a very concise way: "Conscious is becoming, being is unconscious".[53]

This new and modified definition of consciousness takes up a further aspect of monads, namely their continuous becoming. "Our insight into the becoming of organisms allows us to see our conscious life as a permanent struggle", a struggle against our "primitive will" and its "ingrained desires" which is nothing but the "consciousness as a correlate of the bequest of our ancestors."[54] We are a part of the evolutionary process in which "each single individual life is nothing but a bang of a chisel on the forever unfinished statue of our species."[55] Each step, therefore, is "self-transcension", i.e., a move beyond the state already reached which - as transformation - enters the realm of consciousness. To paraphrase Schrödinger: Consciousness is a phenomenon of the evolutionary process.[56] Conscious and unselfish self-transcension, as postulated by the categorial imperative as a foundation of ethics, has its roots in the evolution of consciousness! The inorganic, on the other hand, - and this is a further aspect of a monodological conception - is only "an abstraction".[57]

Schrödinger then draws the consequences. For him the distinction between organic and inorganic is not a difference in the qualities of an object, but depends on the "frame of mind of the subject".[58] But only its frame of mind regarding the organic and dynamic (towards becoming which pushes evolution a bit forward in each individual life) [59], only this allows us to understand ourselves as human beings. By contrast, what is common to all "realms of consciousness" is what is usually called the "real external world".[60]

In summary, we can say that our original metaphor, the crystal with its facets represents the following: unique consciousness which presents itself in the manifoldness of individuals has something in common which becomes manifest in the frame of mind of the subject (in other words, one facet of the crystal), as the real external world. The distinction between subject and object, between organic and inorganic, between 'I' and you is but an immanent distinction of consciousness depending on the given frame of mind. Another essential aspect is the inner dynamic of this unity which consists of the I, consciousness and the world. Thus Schrödinger's view can be characterized as an idealistic dynamic monism.

6. MIND AND BODY: THE WORLD AND I

Now that we have analysed the main aspects of Schrödinger's metaphysics, let us use them to re-examine the mind-body problem. Each theory of the 'I' automatically poses a question that has hardly been dealt with up to now, i.e., how mind and body are to be understood and how the mind (or the soul) and the world interact. This is something Schrödinger is aware of. Repeatedly he describes the paradoxes of free will when mind and body are seen as a duality.[61] For him, this paradox is a consequence of the scientific world view which in accepting only causal or statistical necessity leaves no space for consciousness to intervene. [62] For Schrödinger, the materialistic approaches are just as unsatisfactory as the monistic ones of Spinoza and Haeckel: they do not enable us to solve the old problem of an influxus physicus or to transform into a causal connection. This is the conundrum that results when consciousness is linked as a thinking subject with a body which is understood as an evolutionary process: wouldn't we then have to explain how consciousness could result from a thickening of nerves in the brain as a product of evolution.[63]

Schrödinger's monism tries to overcome the difficulties in basing reality on perceptions of the mind. This is only possible if it is accepted that existence can only be meaningfully attributed to perceptions ,[64] so that correctly speaking, the external world, and with it my body, is nothing but an *idea* of the world (he speaks of "Weltvorstellung"). We have to renounce the material substrate, especially if we intend to understand the processes going on in the body on the basis of the law of entropy and without taking recourse to the entelechies of a vis viva.[65] If I see my body as an idea, as something given in consciousness, then the old mind-body problem is indeed solved. It disappears as metaphysical problems dissolve in Wittgenstein's sense.

Returning to our initial question, i.e., what am I?, we should now try to explain why it is at all possible to speak of a personal 'I'. This question is crucial in view of the fact that Schrödinger's monistic ontology has neither the difference between substance and accidens nor that of subject and object.[66] Instead, he proposes a monadology with only one monad:[67] There is no manifoldness, "in reality all are only aspects of the *unique essence*."[68] Which aspect then constitutes *my* 'I', am 'I' allowed to speak of such - or must ultimately everything become part of the undifferentiated unity of the all embracing consciousness of the unique monad? Schrödinger answered this question by simply quoting Shankara: "'I' and non-'I', within the realm of intuition, are essentially in opposition to each other, ... so that one cannot take the place of the other which implies that their attributes cannot be interchanged." For Shankara - and for Schrödinger as well - it thus follows that "the subjective composed of thinking" is totally in opposition to the "the objective which is perceived as the non-I."[69] In accepting this interpretation of the Vedanta-sutras, Schrödinger emphasizes that "the I - or the mind - can never be an object of research in the strict sense" for all knowledge is already located in it![70] Natural language as well as our science, nevertheless, impute and presuppose at every moment that form of objectivation.[71] From this we clearly recognize that to speak of an 'I' presupposes the distinction between 'I' and 'non-I'; all of Fichte's "Wissenschaftslehre" is based on this insight. The 'I', therefore, is not merely an "accumulation" of "experiences and memories," but "the canvas on which these are fixed."[72]

To employ a metaphor - and this takes us back to the beginning - is compelling: philosophy in this case can only give us a lead. Even to locate a person in a body is just a symbolic act - serving only "practical purposes".[73] Indeed our perceiving and thinking 'I' "is to be found nowhere *in* the world view for it *is* the world view. It is identical with the totality and therefore cannot be in it as a part."[74] It makes no sense to say the world is so and so *before* all consciousness, for "the world only exists *once*...The world as extended in space and time is our imagination."[75]

7. THE 'I' AND MORAL LAW

The idealistic-dynamic monism of Schrödinger goes beyond Aristotle's object-property scheme and even Hume's alternative, i.e., understanding an object as a bundle of properties. By emphasizing the changes within a line of ancestors and including evolution and rejecting the classical concept of the individual, Schrödinger chose a way which can be found in Whitehead's metaphysics of process as well as in the ontology of events, as proposed, for instance, by G. Abel.[76]

It is remarkable that biologists, too, (e.g., M. T. Ghgiselin[77]) have advocated a radical substitution of the old concept of the individual in favor of the concept of species, since individuals as unique living beings are neither separable in each case nor can they fulfill their essential functions (self-preservation and self-reproduction) in isolation. On the contrary: for this they always need another being of the same kind. In shifting this view of an evolutionary dynamic - at least partially - to the development of mental products as well, Schrödinger anticipated the view held today, namely that theories develop in our consciousness in an evolutionary process of becoming. Thus Schrödinger's ontology anticipates various issues at the center of today's discussion.

Of course, only specific parts of Schrödinger's conception can be seen as topcial. More important, therefore, is the question whether his monism is more than mere curiosity. We first have to recall what Schrödinger saw as being at issue. This does not involve primarily epistemological and methodological reflections (they function for him much more as a means for showing that his views are far less abstruse than what they may appear to be at first glance). Schrödinger was, above all, interested in attributing meaning and value. For him a common sphere of values is possible only in terms of a universal consciousness. Since Schrödinger would not admit a separate ontological sphere of values, he aimed at introducing an ethical substratum comparable with the transcendental subject which is in itself just as inalterable as those principles which, in the rationalistic tradition, are characterized as innate. Schrödinger thus believed that the categorial imperative has no absolute grounding while at the same time it is indispensable: "The moral law is simply there, it is universally accepted."[78] He combined this with an evolutionary way of thinking in which norms and values are also subject to evolution. It is thus possible that "a virtue turns into a vice to be rejected."[79] We already saw that for Schrödinger self-transcension is the essential element of evolution. Correspondingly, he sees moral law as being the starting point of a "biological transformation" of man from an egoistic to an altruistic attitude[80] and this biological process of evolution corresponds, as we have seen, to the further development of consciousness. One should not misinterpret this as a biologism, as an attempt to found ethics on biology. On the contrary, moral law is visible on the level of consciousness as an aspect of what we encounter on the level of an objectivistic biology as a special form of self-transcension.

Schrödinger thus proposed a new perspective in ethics as well, a proposal worth pursuing further, as it could eliminate all the naturalistic pitfalls of evolutionistic ethics without being compelled to negate the possibility of ethics at all on the one hand or the acceptability of an evolutionary view on the other.

8. CONCLUSIONS

The process of the development of consciousness must include our own view of evolution, too, so that Schrödinger's metaphysical interpretation of the relation between the world and the 'I' must be integrated in this process of becoming as part of the history of consciousness: although this is never mentioned by Schrödinger himself, he would have accepted it. If we were to pursue this aspect further, Schrödinger's approach would be in keeping with the historical-hermeneutic interpretation of our self-understanding. But at the same time we must admit that even the evolutionary view cannot constitute or lay ground for a world which would (in the order of time as well as in the order of systematic reasoning), precede our idea of the world. This is what Schrödinger bears in mind with regard to the objectivism of the evolutionary view of science. His monism thus brings together two thousand years of occidental scientific tradition with a philosophy of consciousness which has been the central force behind thinking since the 16th century. However, it also enables us to deal with very topical issues which had no common ground until now.

We would misunderstand Schrödinger altogether if we were to interpret the entity of consciousness as if the 'I' were to disappear altogether - as this is essential for the taoistic approach endorsed by Capra. That it is the 'I' who perceives and thinks is the inevitable point of departure which must be taken seriously. What we must give up is the substantiation of the 'I' which - as Leibniz thematized in his monadology - automatically leads to a complete isolation of the individual. Instead, we must acknowledge that we have abilities allowing us to constitute an external world in our consciousness as different from mine and yours, and connected with me by a community through biological and cultural evolution. But we need not substantialize this external world, as is it only an abstraction. In this understanding my 'I' remains an 'I' acting freely and responsibly, which in his thoughts and deeds (through which the 'I' is connected with everything else), develops values, elaborates norms and ultimately grasps the question of meaning and tries to answer it.

Even more significant are the parallels with Capra. It is not necessary to give a full account of Capra's ideas, bookstores are full of books on New Age advocating a synthesis of scientific rationality with an oriental intuitive view of the world as a unity of nature including consciousness. Capra himself does not believe in the possibility of such a synthesis but he speaks of complementation. One may put all this aside, but in his *Aufbau der Physik* von Weizsäcker, claims that Capra and his supporters are "probably a pack on the right track."[81] The success of the New Age movement is its promise of being able to overcome that desideratum which Schrödinger indicated, i.e., the inability of science to say anything about the meaning of life and the universal connections between the I and the world - and this in a time in which this world and with it the species homo sapiens is in danger of being totally destroyed by ourselves as a consequence of our scientifico-technological rationality! Both Capra and Schrödinger refer to the Far Eastern thinking, to models of the unity of the world and 'I'. But Schrödinger is not the beginning of the New Age, for there is a fundamental difference: He *argues* for this unity whereas Capra emphasizes the impossibility of finding a way out by means of rational argumentation. Even if both agree that absolute and final foundations are impossible, Schrödinger insists that we must seek the postulated "highly desirable completion" of our world view in the spirit of occidental rationality, if it is to serve science as an outcome of this tradition, and help us to orient ouselves in the world.

Prigogine begins one of his writings by quoting Schrödinger: "There is a tendency to forget that all of science is linked with human culture as such." The prescientific world view is part of this culture. In this paper I tried to illustrate its reflective penetration and its development with respect to Schrödinger. Schrödinger's proposal for a solution is the argumentative counterpart of the new esotherics. Maybe his monism is too incisive. But here we can clearly say which presuppositions we are not willing to accept: the radical point of departure being the 'I', the unique unity of the world and the 'I', or an extended concept of consciousness - as a hypothesis of revisable metaphysics it is open for criticism. One can thus hardly accuse such a metaphysician of searching for a non-existing cat in a dark room. If we ask a physicist for his solution of Schrödinger's cat paradox, he will become a metaphysicist. But this only becomes a really dangerous situation if someone claims: I have caught the black cat!!

Whether it is really possible to integrate Schrödinger's position into today's scientifico-technological world view is doubtful. But one should understand Schrödinger's attempt to combine occidental scientific rationality with elements of the philosophy of the Upanishads as pointing to a desideratum we must take seriously, namely the lack of meaning in life. He tried to counter this with his integrity and his attempt to base his world view on ethical reflections. To follow his example is still imperative today.

LITERATURE USED:

The quotations stem from the following editions of Erwin Schrödinger's writings:

Geist und Materie, Braunschweig 1959.
Der Geist der Naturwissenschaften, in: Gesammelte Abhandlungen, vol. IV, Allgemein wissenschaftliche und populäre Aufsätze, Vienna 1984.
Mein Leben, meine Weltansicht, Vienna 1985.
Meine Weltansicht, Frankfurt/M. 1963.
Naturwissenschaft und Ethik, in: Gesammelte Abhandlungen, see above.
Naturwissenschaft und Humanismus, Vienna 1951.
Die Natur und die Griechen. Kosmos und Physik, Hamburg 1956.
Was ist ein Naturgesetz? Beiträge zum naturwissenschaftlichen Weltbild, Munich 1962.
Was ist Leben?, Darmstadt 1957.

NOTES

1 *Meine Weltansicht*, pp. 13-14.
2 loc. cit., p. 15.
3 loc. cit., p. 20.
4 loc. cit., p. 26.
5 loc. cit., p. 35.
6 loc. cit., pp. 9-10.
7 Citing E. Cassirer's arguments, Schrödinger rejects P. Jordan's attempt to found free will on quantum theory; cf. *Naturwissenschaft und Humanismus*, p. 81.
8 Was ist Leben?, p. 149.
9 *Meine Weltansicht*, pp. 26-27.
10 loc. cit., p. 27.
11 *Was ist Leben*?, p. 149, pp. 126 ff.
12 loc. cit., p. 27.
13 loc. cit., p. 30.
14 loc. cit., p. 32.
15 loc. cit., p. 33.
16 loc. cit., p. 34.
17 loc. cit., p. 38.
18 loc. cit., p. 47; *Was ist ein Naturgesetz*?, p. 74 and: *Geist und Materie*, p. 40.
19 *Meine Weltansicht*, p. 38 and *Geist und Materie*, p. 91.
20 *Geist und Materie*, p. 40.
21 *Meine Weltansicht*, p. 38.
22 *Meine Weltansicht*, pp. 34-35 f., pp. 131-132; *Was ist Leben*? p.152; *Meine Weltansicht*, p. 38. - Schrödinger discovered the Upanishads in 1918 while lecturing on Schopenhauer (*Mein Leben*, p.18). The texts which Schrödinger uses are: *Sechzig Upanishads des Veda*, translated by P. Deussen, Leipzig 1921. *Die Geheimlehre des Veda*. Ausgewählte Texte, Leipzig 1919 (*Meine Weltansicht*, p.132), and F. Max Müller, *Three Lectures on the Vedanta Philosophy*, London 1894 (*Geist der Naturwissenschaft* IV, p. 379.)
23 C. Fr. von Weizsäcker, *Aufbau der Physik*, Munich 1986, p.604. and cp. 11.2; see p. 592.
24 *Geist und Materie*, p. 1; *Was ist ein Naturgesetz*?, p. 74.
25 *Was ist ein Naturgesetz*?, pp. 74-79.
26 *Meine Weltansicht*, p. 40.
27 *Was ist ein Naturgesetz*?, p. 84.
28 *Meine Weltansicht*, p. 84.
29 *Naturwissenschaft und Ethik* IV, p. 330.
30 *Meine Weltansicht*, p. 44.
31 loc. cit., p. 46.
32 loc. cit., p. 47.
33 loc. cit., p. 53.
34 loc. cit., p. 54.
35 loc. cit., p. 55.
36 *Was ist Leben*?. p. 152.
37 *Meine Weltansicht*, p. 58; *Geist und Materie*, p.1.
38 *Was ist ein Naturgesetz*?, p. 26; cf. p. 63.
39 loc. cit., p, 55.
40 loc. cit., p. 68.
41 *Geist und Materie*, p. 28.
42 loc. cit., p. 28 ff.
43 loc. cit., p. 38.
44 *Naturwissenschaft und Humanismus*, pp. 71-72.
45 *Meine Weltansicht*, p. 59.

46 *Geist und Materie*, p. 1.
47 *Meine Weltansicht*, p. 61, *Geist und Materie*, p. 2.
48 *Meine Weltansicht*, pp. 61-62.
49 *Geist und Materie*, p. 2.
50 *Meine Weltansicht*, p. 63.
51 loc. cit., p. 75.
52 *Geist und Materie*, p. 3, cf. *Meine Weltansicht*, p. 69.
53 *Meine Weltansicht*, p. 76, *Geist und Materie*, p. 6.
54 *Geist und Materie*, p. 7.
55 loc. cit., p. 7.
56 loc. cit., p. 8.
57 *Meine Weltansicht*, p. 63.
58 loc. cit., p. 66.
59 *Meine Weltansicht*, p. 79.
60 *Geist und Materie*, p. 39.
61 *Was ist ein Naturgesetz?*, pp. 68 ff., *Was ist Leben*?, p. 148 ff.
62 *Was ist ein Naturgesetz*?, p. 69.
63 loc. cit., pp. 79-81.
64 *Meine Weltansicht*, p. 93.
65 loc. cit., p. 96.
66 loc. cit., pp. 107-108.
67 loc. cit., p. 131.
68 loc. cit., p. 138.
69 *Der Geist der Naturwissenschaft* IV, p. 379.
70 loc. cit., Iv.381. On Shankara cf. H. von Glasenapp: *Indische Geisteswelt*, vol. I, Hanau 1986, pp. 192-193.
71 *Was ist ein Naturgesetz*?, pp. 34 ff., *Die Natur und die Griechen*, pp. 119 ff., *Geist und Materie*, pp. 27 ff.
72 *Was ist Leben*?, p. 153.
73 *Was ist ein Naturgesetz*?, p. 71.
74 loc. cit., p. 72.
75 loc. cit., p. 81.
76 *Einzelding- und Ereignisontologie*, in: H. Poser, H.-W.Schütt (eds.), *Ontologie und Wissenschaft* (=TUB Dokumentation Kongresse und Tagungen 19), Berlin 1984, pp. 21-50. See the writings of Davidson and Quinton cited in this volume.
77 *Individuality, History and Laws of Nature in Biology*, Lecture at the Wissenschaftskolleg zu Berlin, October 1987.
78 *Geist und Materie*, p. 9.
79 *Meine Weltansicht*, p. 82.
80 *Geist und Materie*, p. 10.
81 Weizsäcker, *Aufbau der Physik*, p. 638.

REMEMBERING SCHRÖDINGER

LEOPOLD HALPERN
Department of Physics, Florida State University
Tallahassee, Florida 32306, USA

When I speak about Schrödinger it almost seems as if I were worshiping him as a hero. The truth is that Schrödinger is one of my intellectual mentors. He did not just influence me through his writings but also as a person. In remembering him, however, I would like to try avoid making a hero out of him. These anecdotes should help the reader get to know Schrödinger not just from the intellectual, but also from the human side. Even in his writings Schrödinger's personality is reflected: he went to great pains in his writings to present a topic as clearly as possible, he commemorated his teachers and documented the source of every idea.

When I think of amusing incidents I again run the risk of seeing him as a hero. Schrödinger's ability to illuminate a situation with striking comparisons was often manifest in his poignant humor.

I was rather naïve about academic life when I assumed my position at the Institute of Physics at the University of Vienna. I had rather unconventional ideas about publishing scientific work at the time: I assumed that a publication is only justified when one has made a new discovery. Schrödinger, although he himself sympathized with such a view, brought up the issue of publication after I had spent one year with him. He was worried about what would become of my career once he became emeritus professor. He once asked me how many publications my research was to yield. Somewhat dismayed I replied: "So far I do not have any yet." "Well, that is not very much," he said. I defended myself by saying that I was dealing with a significant problem which would be worth all the time spent. Schrödinger said, "that reminds me of a colleague who went on a tiger hunt in India and did not kill a single tiger. On his return, somebody remarked that this was not very many, and he replied: "But after all, they were tigers." Schrödinger then said that he personally preferred people who publish little over those who publish too much.

A younger physicist who exhibited considerable self-confidence visited the Vienna Institute of Physics when a successor was being sought for Hans Thirring's chair. Schrödinger did not share the general enthusiasm for the candidate; he found that the latter's simplified physical models and obstruse books reminded him of Surrealism - an artistic movement he was not particularly fond of. This comparison was so fitting that the experienced colleagues spontaneously broke out in laughter when they heard about it. When I visited Schrödinger at his home at that time he made me stay until an unusually late hour. I was beginning to feel uneasy and remarked that I still had to prepare for the next day's seminar, which, due to the lack of more qualified persons, I was in charge of. Schrödinger had always shown understanding for such arguments - but not this evening. I finally said exasperately that I would have to say at least something the next day. "Just stay and don't worry," Schrödinger said, "the candidate visiting our institute will do all of this for

J. Götschl (ed.), Erwin Schrödinger's World View, 169–171.

you." This was also precisely what happened. After a student had finished his presentation no one else had a chance to say anything, even if one or the other correction seemed called for. Schrödinger himself did not come to the Seminar this day.

Schrödinger's humor frequently erupted when he felt somewhat angry. At such moments he became rather moody. One day he angrily showed me a letter in Cyrillic script which he had received from an institute in Russia. He was fluent in twelve languages yet he was unable to decipher this text. "I do not think that the research at that institute will yield anything of much value," he said. When I asked him why he pointed to the letter: "Just have a look at this!" I remarked that at this time research in Russia was cut off from the west but that they had a huge program of their own. "You don't bring about progress through a program," he replied angrily. I agreed but argued that it could perhaps be useful in helping talents to develop. In order to appease him I changed the subject and asked him about the origin of the Cyrillic script. "In school I learned that Saint Cyrill introduced it", he said. "How's that?" I asked. "Well because he was stupid - a saint!" After savoring my amusement he offered a profound explanation. "Cyrill wanted to provide a better means of expression to the richness of the Slavic soul. The Russian language, for example, is so rich in expressions that cannot be translated literally into German - just as it is difficult to translate German into French."

Schrödinger regarded Wolfgang Pauli as a good friend, but he was very critical of his views as a physicist. Most people would have expected the younger and more robust Pauli to survive the ailing Schrödinger. On the day he learned of Pauli's death, Schrödinger only had good things to say, even about Pauli's physics. "I don't understand why he was never elected to membership in the Austrian Academy of Sciences - I wonder if the reason might perhaps be sought in anti-semitism?" he said. I mentioned that Pauli had been openly critical of Austrian physicists, especially during and after the war. At this moment Schrödinger began to pull vigorously on his pipe, a clear sign that a definitive statement would be forthcoming. At last he said: "Pauli was a very honest person."

Schrödinger was the only Nobel Laureate to return to Austria after the war. He was given a small office at the Viennese Institute. This room could be accessed through three doors, which meant that other people often walked through it. One of his few requests was a blackboard. The administration in charge, however, was in no hurry to fulfill this request. I made several futile attempts to accelerate the matter but nothing happened until two days after he had retired, to become professor emeritus. All of a sudden several men appeared with a large blackboard. They hammered a long nail into the wall so that on the adjacent side, in the office of the institute chairman, some of the wall was chipped. Our lab technician became furious but Schrödinger remained calm and only theorized on how government institutions had deteriorated in such a fashion. His interpretation, however, was somewhat incomplete since I later found out that the people employed there were arrested because they had used the new photography equipment to reproduce pornographic pictures.

I once encountered Schrödinger in a state of fury. He showed me a copy of a magazine in which a philosopher had reviewed Schrödinger's books, ascribing his own philosophical views to Schrödinger. Schrödinger cancelled the subscription to the journal (which was hardly responsible for what had happened), but did not react to the author. If something written did not seem lucid enough he often reacted angrily by throwing it away and leaving it at that.

Erwin Schrödinger never recommended his books and was equally critical of his own writings and those of others. He once surprised me with his opinion regarding a long-standing textbook: "This book is miserable. To be sure, the author is competent and profound, but it reminds me of the legislation during the monarchy. Something is first defined and described in a convoluted way. What finally becomes clear is then forbidden."

Schrödinger never seemed satisfied with the results of his research. It appeared to me as if his mind found no rest until he had discovered an unsolvable problem. This may also apply to many situations in life. A most critical moment for me was when the head of the institute in Vienna had literally forced me to work on a project while I was doing research abroad which I was not interested in. When I informed him about this after he had become professor emeritus he replied: "Forcing someone to do a specific research project is something I have never come across.[1] I am worried about your future; since you have been asked to do this by the head of your own institute, I cannot give you any other advice but to comply. "However," he added, "you will see that it is impossible to do research on something that does not appeal to you!"

Schrödinger had a weak constitution but iron will power. This became manifest when physicians had given up on him. When I came to bid him farewell, I found him wrestling with scientific and philosophical problems. This helped him again and again to recover. It was particularly impressive how he tried to teach and hand down to us as much as possible, in spite of the difficulties he had speaking. At such moments his mind appeared superior than ever before.

NOTES

1 Coercion is something that did not exist for him. "This is not a religion class and so you do not have to believe anything," he once said during a discussion.

THE STUDY OF NATURE AND THE EMERGENCE OF WORLD VIEW - PHILOSOPHICAL REFLECTIONS ON E. SCHRÖDINGER'S APPROACH

JOHANN GÖTSCHL
Ludwig Boltzmann Institut für Wissenschaftsforschung und
Institut für Philosophie der Universität Graz
Mozartgasse 14, 8010 Graz, Austria

The great Greek natural philosophers, who Erwin Schrödinger turned to when dealing with specific issues, provided man with a new approach to nature.[1] The study of nature and man requires a structured approach based on specific principles which today are best described as "rational thinking" or scientific thinking.

Modern philosophy has been under the sway of various insights of Greek philosophy (in particular, Greek natural philosophy) which have been reformulated and transformed. Nevertheless, it seems to be characteristic of all modern thought that our world view and our knowledge of nature no longer form a unity.[2] This is one of Schrödinger's central insights. He tried to identify the role of knowledge of nature in the Greek philosophy as well as in 20th century physics.[3] In keeping with the basic orientation of modern European thought Schrödinger realized that world view implies much more than knowledge of nature.[4]

The former is more general, referring not only to nature but to both nature and man. World view always ultimately seeks to grasp the totality of human existence, the position of man in nature and culture in a larger context, the relationship between the universal and the singular. According to E. Schrödinger,[5] world view is the conceptual framework for understanding the development or evolution of man and nature as a synthesis.

Ever since R. Descartes divided the world into two spheres *(res extensa* and *res cogitans)*, it has become increasingly clear that the study of nature differs from that of culture. And ever since physics and biology obtained a final breakthrough in culture, ushering in paradigmatic innovations, the European understanding of man has taken on a new dimension.[6] The orientation of man in the 20th century has become a new paradigm that implies the following: the role of the study of nature, the postulated outer world and in part the other world counterposed by man are constitutive for a world view befitting this day and age.

This is the novel imperative of modernism, in particular, of the 20th century, around which Schrödinger's philosophical thinking revolves.[7] Schrödinger struggled to develop a philosophical understanding of the world in which science is an integral part: for him world view is based on scientific ideas incorporated into or understood as philosophy.[8] Up to the present the study of nature aims at an understanding the world - this, at least, is the tacit assumption - in which nature is the dominant feature. This manifests a view of man which is "accurate," "true," but above all "free of dogmas."[9] The implicit imperative of thinking in the 20th century is to base our world view on our knowledge of nature, since man is part of nature. What we know from nature, we know from science.[10]

The dominant role played by our knowledge of nature in the formation of our world view, i.e., in the evolution of a synthesis of nature and man, has, in this century, prompted some people

J. Götschl (ed.), Erwin Schrödinger's World View, 173–181.

to believe that the Cartesian split of the world can be superseded by highly complex reductionist models.[11] Accordingly, this new modern view of the world has to have scientific underpinnings, i.e., a scientific representation of the world.[12] The increase in scientific knowledge tends to confirm the insight that man is nothing more and nothing less than a highly differentiated concretization of nature.[13]

The growing importance of science points to its cultural or existential relevance that transcends all cultural differences. Science becomes constitutive for the idea of an objective that is compatible with the unity of scientific knowledge and the diversity of cultures.[14]

This would imply higher levels of interhuman or intercultural understanding as a product of a process of self-organization.[15] It is not surprising that in the 19th century, and even more so in the 20th century, scientific knowledge is seen as man's optimum realization of himself on the basis of scientifc theory. This scientific knowledge does not just reflect knowledge of nature, it also shapes man's dynamic relation to nature. Science as a specific form of knowledge would contribute most to forming a certain world view. This idea can be expressed as the following equation: The increase in knowledge of nature corresponds to greater human self-understanding.[16] This, at least, is the tacit assumption of modern science from which also E. Schrödinger[17] develops his philosophy or cosmology. At this point we might ask how Schrödinger sees the relation of nature and what role he ascribes to the "study of nature" in the construction or the evolution of a modern world view. Thus we will attempt in the following to reflect on Schrödinger's approach from a philosophical perspective.

According to Schrödinger, the role of science in culture becomes evident when the scientific-rational patterns of thought are examined.[18] Science was, and still is, the most powerful weapon against any sort of in the guise of metaphysics. The major world-view models, e.g., Leibniz, Schopenhauer, Nietzsche, etc., illustrate best what can be understood as metaphysics. It is decisive that science tends to eliminate metaphysics, or at least to make it superfluous - this, at least, is what positivists in the strict sense hold.

Schrödinger, however, claims that a philosophical reflection of the relationship between scientific and metaphysical knowledge requires a more differentiated answer than the positivist one. The positivists failed to identify the possible function of metaphysics in the accrual of knowledge of nature.[19] Schrödinger attributes two functions to metaphysics which puts him in the vicinity of thinkers of the 20th century such as E. Cassirer[20], N. Whitehead[21] and K. R. Popper.[22]

One of Schrödinger's central ideas is that metaphysics plays a crucial role for the accrual of our knowledge of nature. Man's creativity does not just manifest itself as a rational construct or as an algorithm. This position bears resemblance to Popper's theory of science. K. R. Popper showed that metaphysics can have a cognitive-heuristic function but only in a context of discovery - not in a context of justification. The latter is purely deductive. In trying to define the relationship between science and metaphysics as well as its importance for our world view, Schrödinger goes beyond this. For Schrödinger, metaphysics plays an equally crucial role in scientific knowledge, a process of understanding which brings forth representations of the world.[23] Scientific knowledge is integrated into a cultural network through understanding, and it is here that we always encounter metaphysics. But Schrödinger does not see science as metaphysics. Science would probably degenerate to or bring forth - if anything at all - purely ratio-technical forms of thought and civilization.[24]

If we follow Schrödinger in assuming that world view is a synthesis of knowledge of nature and human self-understanding, it then becomes clear to us why our world view is impossible without science but at the same time cannot be fully explained by science. He thus sees a purely scientific view of the world as being impossible. Schrödinger provided the basic insights into the role of the knowledge of nature in our formation of world view. In essence, these are two essential aspects.

1. Scientific knowledge is characterized by invariance, which is most pronounced in physics. The role of this invariance for the formation of world view is basically as follows: invariants are constructive principles which lead to a greater differentiation of subject-object relations. The culmination of this process of differentiation is the theoretical elimination of the cognizing subject.[25] In the invariants of physics, this relational construct between subject and object (man and nature) which gives objectivity and truth their semantic meaning becomes explicit. The creation of invariants transforms subjectivity as a central source of viewing the world on various levels of knowledge of scientific objectivity. In this process the way nature relates to the lifeworld is transformed through physics. On the one hand, subjectivity is a source of knowledge and on the other hand, invariance is a principle of structuration and selection central to the formation of our view of the world.[26]

2. The further scientific knowledge progresses, the more evident it becomes that nature and culture create a dynamic unity. Schrödinger shows that the function of science as a representation of the world reveals the following: the higher the level of science, the better our understanding of man as an integral part of nature without reductionism. The reason for this is that scientifc progress has made a greater number of scientific self-referential theories possible. On this basis, man becomes part of nature in a non-reductionist sense.[27] This integration of scientific knowledge in the cultural process of consciousness is culture in the sense of a transformed reflection of nature. This integrative way of thinking is a source of scientific theories thanks to its innovative subjective character. From Schrödinger's philosophical perspective physics seems to be dependent on a general understanding of the world. This, however, does not imply that physics has to be interpreted in a cultural-relativistic sense.[28] Rather, Schrödinger sought to make us aware of the fact that metaphysics is an immanent part of the lifeworld in both the creation and the foundation of physics as a product of culture. On this issue, there is an interesting affinity between Schrödinger and Whitehead. Both models of a cosmological world view attempt to explain the hypothesis that the development of man can only be understood in terms of science, because development can only be understood as an irreversible and non-linear process.[29] Schrödinger points our attention to the fact that science and metaphysics can complement each other and that science is not the only approach to nature.[30] Schrödinger calls upon our good judgement not to replace an earlier unreflected metaphysics with a unreflected physics of the 20th century.

Schrödinger's tolerance for metaphysics is very differentiated. He shows that the various metaphysicians who conceived of world views as integral models of the cosmos and life also developed highly complex scenarios. These scenarios were analytically influenced by the thought categories of a scientific-technical calculus, which resulted in new scientific insights. To Schrödinger, the transcension principle inherent in all metaphyics and science is central to metaphysics and science and independent of specific content.[31] It is a principle by virtue of which one can transcend the specific singularities of both nature as well as culture. This is much more than scientific universalism. Metaphysics has the function of being an innovative cultural model for scientific knowledge. In this sense, science is inseparably linked to metaphysics in the "transcension principle." For Schrödinger, metaphysics is a manifestation of this principle. It is logos from which interest in knowledge can become constituted. Metaphysics, however, is neither just prescientific nor interpretative form of knowledge. According to him, science can be understood as that part of general logos through which it is possible to represent the infinite diversity of singularities of the material world in conceptual frameworks. This is the essence of Schrödinger's view of the relation between physics and metaphysics. Metaphysics is not - as is often postulated - the possible precursors of physics. In this understanding physics would be primarily the result of empirical-experimental metaphysics selected. Schrödinger does not aim at reinstating metaphysics but at showing that metaphysics is an immanent part of scientific theory formation in which physics reveals its function as a representation of world. Metaphysics and physics form a continuum in the process of knowledge. The fact that metaphysics does show itself in natural scientific hypotheses and theories as a result of knowledge, has little to say about the possible metaphysical meaning of physics. This metaphysical meaning only becomes visible in the reconstruction of the relations between physics and the surrounding context of physics, especially when the content of scientific theories is reconstructed in its philosophical-existential meaning. The result is the following: nature and man, object and subject merge in our view of the world.[32] The transcension also inherent in traditional physics and metaphysics explains how object and subject originate from the same source; even though they are not identical they relate to each other in an evolutionary way.

Schrödinger goes further in his thinking, as we said above. He helps us to understand the difficult question as to how the study of nature relates to our view of the world. How did Schrödinger go about answering this question? This issue has to do with the relevance of culture, lifeworld and existence for our understanding of nature. Here world view is understood as a synthesis of our understanding of nature and human self-understanding. This involves general orientations from ethical issues to issues such as death. Schrödinger did not just have original ideas on these issues, he also provided a number of incentives for what is presently being discussed in philosophy and science and for future thought.

It can be first be stated that Schrödinger saw the "subjectivity principle"[33] and the "principle of transcension" as the motor of the human mind. According to Schrödinger, these two principles allow man to define his own specificity. The most significant characteristic of man is his creativity. Or to put it differently: in Schrödinger's approach to world view and the issue of how man relates to nature is not reduced through a static view to a material-complex system and thus neutralized in ethic terms. Rather, man is identified as a being that changes and develops itself and its basic structures by understanding and forming nature. The cognizing subject that brings forth concretizations through the principles of subjectivity and transcension becomes aware, on higher levels of reflection, of its own dynamic and the internal connectedness of this dynamic with the dynamic of nature. This is a crucial step in Schrödinger thinking: it means a structural and material interrelatedness between our understanding of nature and our understanding of ourselves.[34]

Schrödinger's ideas on the "the study of nature and world view" lead him even further. K. R. Popper and others have shown in the last decades, in contrast to to the positivism of the 20th century, that metaphysics plays a heuristic function at least in the process of scientific understanding but not in its foundation. According to Popper and other adherents of critical rationalism: scientific foundations follow a strictly logically structured algorithm, and in this respect metaphysics plays no role in the foundation of knowledge. Schrödinger, however, differentiates much further on this issue. Like E. Cassirer and A. N. Whitehead, Schrödinger sees the relation between creation and foundation of knowledge as being much more holistic. He assumes that the foundation of a model of knowledge, e.g., the central idea of a scientific theory, is always already partially inherent in the model of knowledge. Schrödinger's interpretation of cognitive models can be analysed in the following central dimensions:

(1) cognitive models are conceived within the context of a world view. The models are totalities of ideas; their differentiation in terms of a scientific discipline is a process that takes place later.[35]

(2) a world view cannot be based only on scientific models of cognition, since the principles and rules by means of which scientific knowledge takes place do not appear in the natural sciences.[36]

(3) All cognitive models - in particular those of scientific knowledge are an integral part of culture - since it is not possible to make a clear distinction between what science is and what it is not.[37]

With this differentiated view of Schrödinger's thinking, we come a bit closer to answering our original question as to the role of science for our view of the world. We can now state the following: the categorial distinction between object and subject, matter and mind can no longer be maintained when philosophical considerations come to bear in science. The "cognizing" active subject is no longer just part of the world as something that cannot be homogenized with the world. The subject represents the world as "world itself", as Schrödinger often says in a heuristic sense.

This is, no doubt, a central, perhaps the most central aspect of Schrödinger's philosophical thinking, even if it is something he least pursued. This aspect touches the whole epistemological and world-view implications of the study of nature, when it is viewed in the context of the current debate on the relation of man and nature. This seemingly simple idea according to which man is not part of the world but world himself, world as a whole, shows clearly that Schrödinger was aiming at a cosmological model of the world that could not just be reduced to physics or physical cosmology. It is, in itself, an approach to philosophical cosmology or natural philosophy, in which the "cognizing" subject is represented as a consistent and coherent element. Here Schrödinger suggests the prerequisites of a homogenization of nature and man, matter and mind, brain and consciousness.[38]

Schrödinger seeks a unity, that is, the dynamic unity of man and nature understood as a process. This is exactly what current attempts at a holistic interpretation of nature and man aim at. A holistic understanding of man ultimately means that nature and man can be represented consistently with the notion of "world." There is no doubt that what Schrödinger had in mind was a construction of the world that corresponds to today's views of the world in the theories of self-organization or chaos.[39] Like thinkers before Schrödinger, he also tries to solve the paradox of the cognizing subject being interpreted in the context of the object, as can be seen in science. Neither I. Kant nor A. N. Whitehead were able to use their huge conceptual constructions to solve this paradox in a satisfactory way. How does Schrödinger go about solving this problem?

Recent theories of self-organization and chaos have shown that it is now possible to approach a unified or holistic model. These theories are theories of nature (matter) but are also capable of providing an integral representation of the relation of nature and man. This is precisely what Schrödinger had in mind. He tried to show that the cognizing and active subject can only be understood in terms of man's interaction with non-human nature. Human nature is that part of nature through which the transformation of nature is a necessity and not just as a possibility. In this process Schrödinger sees a foundation for his theoretical approach to nature and man. Such an approach would make it possible to eliminate the classical-categorial separation of science and ethics. This insight into the necessity of interaction through which man and nature are transformed points to the limits and possibilities of science for our understanding of the world. According to Schrödinger man and nature must stem from the same principles. Increased scientific knowledge of nature and man provide evidence that our knowledge of the outer world is isomorphic with our knowledge of man's inner world. This enables us to understand what Schrödinger tried to explain by means of two theorems:

(1) Man is neither part of the world, nor is he just a product of nature. Rather, man represents nature.[40]

(This is Schrödinger's holistic approach to man. This allows us to go a step further in determining the role of the study of nature in our view of the world.)

(2) Man and nature do not coexist. Rather, they are interrelated through coevolution; nature and man can only be understood as interlocking systems[41].

From this we see the relevance of the study of nature for the formation of our world view on the basis of scientific methodology. With (1) and (2) Schrödinger laid the groundwork for a philosophy of nature or philosophical cosmology, in which scientific knowledge plays a specific role. Schrödinger saw that a consistent conceptual representation of man and nature is only possible in a new philosophy of nature or philosophical cosmology based on science. With his position Schrödinger renounces any sort of reductionism. The role of science for world view is thus set down: natural science is a necessary, albeit not sufficient foundation for our view of the world.

These premisses for a philosophical cosmology can be expressed in the following way in Schrödinger's approach: (1) Man, a cognizing subject, a dynamic unity of brain and consciousness, as well as the non-human nature described by man, are phenomena of an inherent principle, i.e., the deeper layers of reality. Schrödinger does not see a religious-theological interpretation of the world as necessary. (2) Our understanding of nature through natural science is a medium of self-knowledge. Neither natural science is a representation of the world nor does the latter result in logic-causal succession from science. Rather: the integration of knowledge of nature into the cultural network through understanding and transforming it, forms a representation of the world.[42]

Precisely this process of an " integration of knowledge of nature into the cultural network on the basis of understanding" is what Schrödinger is interested in. He sees ethics, moral and humanity as being more than just peripheral elements of natural science and tries to integrate these elements with knowledge. Schrödinger tries to show that the development of morality in the course of the evolution of man and nature takes place on a separate level of development, but at the same time that it is interrelated with the totality of nature. This also enables Schrödinger to avoid naturalistic fallacies. He clearly sees that:

(1) no moral laws can be deducted from natural laws

(2) morality, ethics and humanity show that a world view containing these elements reveals emerging manifestations of the evolutionary process.

What implications do these insights have for the "study of nature and world view?" Unlike some sociobiologists today, Schrödinger never saw the categories of morality as being simply genetically determined or fixed forms of behavior. He also never tried to secretly deduct ethics and morality from genetic premisses. Rather, Schrödinger addressed issues crucial for any world view: e.g., how can nature and morality be unified; can there be such a thing as a consistent representation?

According to Schrödinger there is no such thing as scientific explanation or reduction of morality, ethics and humanity. He does, however, see the possibilities and limits of natural science contributing to a scientific understanding of morality, ethics and humanity. Here we have to reach the limit of current scientific understanding in terms of morality. This also marks off the area in which natural science can play a role for world view. A reconstruction of Schrödinger's ideas shows to what extent natural scientific methodology and the creation of theories are valid for the cognizing subject. Unreflected application of scientific methodology (categories) results in a break in both our understanding of nature and of ourselves. The cognizing subject is neither a linear nor a non-linear extension of the object of natural science. Here, no doubt, lies a key to understanding the relation between nature and morality. With these reflections, Schrödinger obviously wanted to show a thrust of thought, which he expresses in terms of a heuristic hypothesis: the central issue is to arrive at a "scientific understanding" of morality, morality being understood as a new manifestation in the development of the relations between man and nature. Here Schrödinger draws, at least implicitly, a vision of what is central in : man and nature are to be viewed as forming a unity so that apparent inconsistencies can be overcome.

On this issue, Schrödinger is most interested in identifying the structural and the material interrelationships between nature and culture, matter and mind, brain and consciousness. The highest material and structural structures are those of being and ought. Or to put it differently: Schrödinger asks what price must be paid to obtain a natural-scientific understanding of morality. Obviously Schrödinger was on the right track to the insight that sciences dealing with the non-human nature would have to be extended to comprise the "natural sciences of man" as well. These new emergent sciences have brought about a shift in the old problem of the relationship of mind and matter.[43] This increased knowledge of the natural sciences of man has resulted in a new concept of matter and a transformation of the focus or production of theories. According to Schrödinger, we should not just question the moral function of biological determinants. Rather, we should begin to reflect on the biological function of morality relevant to our view of the world.[44] The advance made by Schrödinger is important in this context, since morality cannot relate to nature as a whole.

Schrödinger must have seen morality, ethics and humanity - understood here as a new entity within the evolutive process dynamics - as a phenomenon proving that the concept of "nature" cannot be characterized only in concepts of natural science, i.e., material structures. Morality also indicates that the new understanding of nature implies that nature is conditioned by culture and that natural history and cultural history form a dynamic unity.[45] Morality - its manifestation, influence and continuity - is constitutive for the relation between man and nature and thus a foundation of all, which according to Schrödinger, can be seen as a process of humanization.

A concomitant phenomenon of this deeper understanding of the relations between man and nature is the ongoing transformation of evolution from egoism to altruism.[46]

NOTES

1 Schrödinger, E., *Nature and the Greeks*, Cambridge 1954.
Schrödinger, E., *Collected Papers on Wave Mechanics*, New York 1978.
Urban, P., "Die Antrittsvorlesung E. Schrödinger in Graz über die 'Grundidee der Wellenmechanik'" 1936," in: *Mitteilungen der Österreichischen Gesellschaft für Naturwissenschaften* 4 (1984).
Schrödinger, E., *Briefe zur Wellenmechanik*, ed. K. Przibram. Vienna 1963.
2 Schrödinger, E., see 1a), esp. chapter VII.
3 Schrödinger, E., see 1a), esp. chapter VII.
4 Schrödinger, E., *Mein Leben, Meine Weltansicht*, Vienna-Hamburg 1985, esp. pp. 47 ff.
5 Schrödinger, E., see 4), esp. pp. 47 ff.
6 Schrödinger, E., see 4), esp. p. 117.
7 Shimony, A, "Reflections on the Philosophy of Bohr, Heisenberg and Schrödinger," in: *Physics, Philosophy and Psychoanalysis*, ed. by R. S. Cohen and L. Laudan, Dordrecht. Reidel 1983, pp. 209-221.
8 Schrödinger, E., see 4), esp. pp. 121-128.
9 Schrödinger, E., *Über Indeterminismus in der Physik. Ist die Naturwissenschaft milieubedingt?* Zwei Vorträge zur Kritik der naturwissenschaftlichen Erkenntnis, Leipzig 1932, esp. pp. 25 ff.
10 Schrödinger, E., see 1a), esp. chapter VII.
11 Olby, R., "Schrödinger's Problem: What is Life?," in: *Journal of the History of Biology 4* (1971), pp. 119-148.
12 Schrödinger, E., *Was ist ein Naturgesetz?* Beiträge zum naturwissenschaftlichen Weltbild, Oldenburg-Munich-Vienna 1967, esp. pp. 27-85.
13 E. Schrödinger's contributions to eliminating the seeming categorial difference between mind and matter, cf. E. Schrödinger, *Mind and Matter*, Cambridge 1958.
14 Schrödinger, E., see 9).
15 Schrödinger, E., see 13).
16 Schrödinger, E., see 13).
17 Schrödinger, E., see 12), esp. chapter IV: "Die Lücken, die aus der Objektivierung entspringen," pp. 59-85.
18 Schrödinger, E., "Der erkenntnistheoretische Wert physikalischer Modellvorstellungen," in: *Jahresberichte des Physikalischen Vereins zu Frankfurt* (1928/1929), pp. 44-51.
19 Even though positivists and empiricists introduced, in the first half of this century, the distinction between "context of justification" and the "context of discovery", the potential role of metaphysics for the "context of discovery" has yet to be identified. cf., e.g., R. Carnap, *Der logische Aufbau der Welt*, 1928; H. Reichenbach, *Experience and Prediction. An Analysis of the Foundations and the Structure of Knowledge*, Chicago 1937.
20 Cassirer, E., *Das Erkenntnisproblem in der Philosophie und Wissenschaft der neuen Zeit*, vol.III, Berlin 1920.
21 Whitehead, A. N., *Process and Reality*. An Essay in Cosmology, London 1977.
22 Popper, K. R., *Logik der Forschung*, Tübingen 1966.
23 Schrödinger, E., see 12), esp. pp. 37-40.
24 Schrödinger, E., see 4), esp. pp. 50 ff.
25 Schrödinger, E., see 9), esp. pp. 54 ff.
26 Schrödinger, E., see 12), esp. pp. 62 ff.
27 Schrödinger, E., see 13), esp. pp. 27-49.
28 Schrödigner, E., see 9), esp. pp. 55 ff.
29 Schrödinger, E., see 13), esp. chapter II.
30 Schrödinger, E., see 1a), esp. chapter I.
31 Schrödinger, E., see 4), esp. pp. 52 ff.
32 Schrödinger, E., see 12), esp. pp. 80 ff.
33 Schrödinger, E., see 9), esp. pp. 35 ff.

34 Schrödinger, E., see 1a).
35 Schrödinger, E., see 4), esp. pp. 47-51.
36 Schrödinger, E., see 13), esp. pp. 60 ff.
37 Schrödinger, E., see 12). cf. esp. E. Schrödinger's analysis of the relation between the hypothesis of understandability and objectivation, pp. 34 ff.
38 Schrödinger, E., see 13), esp. chapter V.
39 Götschl, J., "Philosophical and Scientific Conceptions of Nature and the Place of Responsibility," in: *International Journal for Science Education*, vol.12, no. 3, pp. 288-296.
cf. also J. Götschl, "Die Bedeutung des Paradigmas der Selbstorganisation für Naturverständnis und Selbstverständnis," in: *Selbstorganisation. Aspekte wissenschaftlicher Revolution*, ed. W. Krohn and G. Küppers, Braunschweig- Wiesbaden 1990.
40 Schrödinger, E., see 1a), esp. chapter VII.
41 Schrödinger, E., see 13), esp. chapter II.
42 Schrödinger, E., see 12), esp. chapter II and III.
43 Schrödinger, E., see 12), esp. p. 84 f.
44 Schrödinger, E., see 4), esp. pp. 110-118 and p. 181.
45 Leinfellner, W., "Schrödinger der Physiker: Ordnung gegen Entropie," in: *Zeitschrift für Wissenschaftsforschung* 1991 (in print).
46 Schrödinger, E., see 4), p. 117.

BIBLIOGRAPHY OF ERWIN SCHRÖDINGER'S WORKS

Schrödinger's writings have been compiled in several more or less detailed lists. Among those aiming at completeness there are the bibliographies in "Poggendorff"[1], by E.-E. Koch[2] , by W. T. Scott[3] and of H. Ezawa[4].

The first bibliography based on records that had been the personal property of E. Schrödinger, was compiled for the publication of "Erwin Schrödinger: Gesammelte Abhandlungen. Collected Papers. Wien: Österreichische Akademie der Wissenschaften. Braunschweig: Vieweg. 1984" by A. Dick and G. Kerber. Forming the basis for the work on hand, it was supplemented and updated, providing the reader with a new access to Schrödinger's thought.

Schrödinger's publications are listed in two separate parts. Part A holds articles of any kind, part B his published books, where the usual bibliographical description of the original version is followed by the table of contents. Translations have been mentioned in chronological sequence according to the "index translationum".

Comments and explanations added by the compilers are put in square brackets. Further publications of the same article are indicated by numbers, e.g. [A 110, B 16.1] means: Bibliography part A, No. 110 and part B, No. 16.1. Nearly all articles listed below - but no books - have been reproduced in the four volumes of the aforementioned edition "Erwin Schrödinger: Collected Papers...". Therefore no additional indication regarding a reprint of an article in this publication is given.

---

1 J. C. Poggendorffs biographisch-literarisches Handwörterbuch... Bd. 5,6,7a. Leipzig, Berlin: Verlag Chemie. 1926, 1940, Berlin: Akademie-Verlag. 1960.

2 E. Schrödinger: Die Wellenmechanik. Stuttgart: Battenberg. 1963, (Dokumente der Naturwissenschaften, Abteilung Physik, Bd. 3, ed. by A. Hermann) p. 194-199

3 W. T. Scott: Erwin Schrödinger. An Introduction to his writings. Amherst: Univ. of Massachusetts Press. 1967, p. 151-161

4 H. Ezawa: Hadô rikigaku keiseishi. Shurêdingâ no shokan to shôden. Tôkyô: Misuzu shobô. 1982, p. 259-269 [Bibliography in English, except for the few books published in Japan]

A. ARTICLES

1910

1 *Über die Leitung der Elektrizität auf der Oberfläche von Isolatoren an feuchter Luft*
Sitzungsberichte der kaiserlichen Akademie der Wissenschaften in Wien. Mathematisch-naturwissenschaftliche Klasse, Abteilung 2a, **119**, (1910), 1215-1222

1912

2 *Zur kinetischen Theorie des Magnetismus (Einfluß der Leitungselektronen)*
Sitzungsberichte der kaiserlichen Akademie der Wissenschaften in Wien. Mathematisch-naturwissenschaftliche Klasse, Abteilung 2a, **121**, (1912), 1305-1328

3 *Studien über Kinetik der Dielektrika, den Schmelzpunkt, Pyro- und Piezoelektrizität*
Sitzungsberichte der kaiserlichen Akademie der Wissenschaften in Wien. Mathematisch-naturwissenschaftliche Klasse, Abteilung 2a, **121**, (1912), 1937-1972

4 *Über die Höhenverteilung der durchdringenden atmosphärischen Strahlung (Theorie)*
Sitzungsberichte der kaiserlichen Akademie der Wissenschaften in Wien. Mathematisch-naturwissenschaftliche Klasse, Abteilung 2a, **121**, (1912), 2391-2406

1913

5 *Notiz über die Theorie der anomalen elektrischen Dispersion*
Verhandlungen der Deutschen Physikalischen Gesellschaft, **15**, (1913), 1167-1172

6 *Radium-A-Gehalt der Atmosphäre in Seeham 1913*
Sitzungsberichte der kaiserlichen Akademie der Wissenschaften in Wien. Mathematisch-naturwissenschaftliche Klasse, Abteilung 2a, **122**, (1913), 2023-2067

1914

7 *Über die Schärfe der mit Röntgenstrahlen erzeugten Interferenzbilder*
Physikalische Zeitschrift, **15**, (1914), 79-86

8 *Zur Dynamik elastisch gekoppelter Punktsysteme*
Annalen der Physik, (4), **44**, (1914), 916-934

9 *Zur Theorie des Debyeeffekts*
Physikalische Zeitschrift, **15**, (1914), 497-503

10 *Über die weiche (β) Sekundärstrahlung von γ-Strahlen* (with K. W. F. Kohlrausch)
Sitzungsberichte der kaiserlichen Akademie der Wissenschaften in Wien. Mathematisch-naturwissenschaftliche Klasse, Abteilung 2a, **123**, (1914), 1319-1367

11 *Zur Dynamik der elastischen Punktreihe*
Sitzungsberichte der kaiserlichen Akademie der Wissenschaften in Wien. Mathematisch-naturwissenschaftliche Klasse, Abteilung 2a, **123**, (1914), 1679-1696

12 *Dielektrizität*
Handbuch der Elektrizität und des Magnetismus, I, [1914], 157-231, Leipzig: Barth. 1918.

1915

13 *Notiz über den Kapillardruck in Gasblasen*
Annalen der Physik, (4), **46**, (1915), 413-418

14 *Zur Theorie der Fall- und Steigversuche an Teilchen mit Brownscher Bewegung*
Physikalische Zeitschrift, **16**, (1915), 289-295

1917

15 *Die Ergebnisse der neueren Forschung über Atom- und Molekularwärmen*
Die Naturwissenschaften, **5**, (1917), 537-543

16 *Die Ergebnisse der neueren Forschung über Atom- und Molekularwärmen (Schluß)*
Die Naturwissenschaften, **5**, (1917), 561-567

17 *Zur Akustik der Atmosphäre*
Physikalische Zeitschrift, **18**, (1917), 445-453; Nachtrag, 567

1918

18 *Die Energiekomponenten des Gravitationsfeldes*
Physikalische Zeitschrift, **19**, (1918), 4-7

19 *Über ein Lösungssystem der allgemein kovarianten Gravitationsgleichungen*
Physikalische Zeitschrift, **19**, (1918), 20-22

20 *Über ein in der experimentellen Radiumforschung auftretendes Problem der statistischen Dynamik*
Sitzungsberichte der kaiserlichen Akademie der Wissenschaften in Wien. Mathematisch-naturwissenschaftliche Klasse, Abteilung 2a, **127**, (1918), 237-262

21 *Notiz über die Ordnung in Zufallsreihen*
Physikalische Zeitschrift, **19**, (1918), 218-220

1919

22 *Wahrscheinlichkeitstheoretische Studien, betreffend Schweidler'sche Schwankungen, besonders die Theorie der Meßanordnung*
Sitzungsberichte der kaiserlichen Akademie der Wissenschaften in Wien. Mathematisch-naturwissenschaftliche Klasse, Abteilung 2a, **128**, (1919), 177-237

23 *Der Energieinhalt der Festkörper im Lichte der neueren Forschung*
Physikalische Zeitschrift, **20**, (1919), 420-428; 450-455; 474-480; 497-503; 523-526

24 *Über die Kohärenz in weitgeöffneten Bündeln*
Annalen der Physik, (4), **61**, (1919), 69-86

1920

25 *Theorie der Pigmente von größter Leuchtkraft*
Annalen der Physik, (4), **62**, (1920), 603-622

26 *Grundlinien einer Theorie der Farbenmetrik im Tagessehen*
Annalen der Physik, (4), **63**, (1920), 397-426; 427-456; 481-520

27 *Farbenmetrik*
Zeitschrift für Physik, **1**, (1920), 459-466

1921

27a *Webster, D. L.: Quantum emission phenomena in radiation, Found, C. G.: Ionization potentials of argon, nitrogen, carbon monoxide, helium, hydrogen and mercury and iodine vapors, Wilson, H. A.: On electromagnetic momentum* [Articles Review]
Die Naturwissenschaften, **9**, (1921), 20-21

28 *Versuch zur modellmäßigen Deutung des Terms der scharfen Nebenserien*
Zeitschrift für Physik, **4**, (1921), 347-354

29 *Isotopie und Gibbssches Paradoxon*
Zeitschrift für Physik, **5**, (1921), 163-166

1922

30 *Dopplerprinzip und Bohrsche Frequenzbedingung*
Physikalische Zeitschrift, **23**, (1922), 301-303

31 *Über die spezifische Wärme fester Körper bei hoher Temperatur und über die Quantelung von Schwingungen endlicher Amplitude*
Zeitschrift für Physik, **11**, (1922), 170-176

32 *Über eine bemerkenswerte Eigenschaft der Quantenbahnen eines einzelnen Elektrons*
Zeitschrift für Physik, **12**, (1922), 13-23

1923

33 *Ton und Farbe*
Neue Zürcher Zeitung, 3. Februar, (1923)

33a *Vom Radium*
Neue Zürcher Zeitung, 15. März, (1923)

1924

34 *Gasentartung und freie Weglänge*
Physikalische Zeitschrift, **25**, (1924), 41-45

35 *Kann man Atome photographieren?*
Neue Zürcher Zeitung, 7. Februar, (1924)

36 *Über das thermische Gleichgewicht zwischen Licht- und Schallstrahlen*
Physikalische Zeitschrift, **25**, (1924), 89-94

37 *Bemerkung zu zwei Arbeiten des Herrn Elemér Császár über Strahlungstheorie und spezifische Wärmen*
Zeitschrift für Physik, **25**, (1924), 173-174

38 *Bohrs neue Strahlungshypothese und der Energiesatz*
Die Naturwissenschaften, **12**, (1924), 720-724

39 *Einiges über die Sterne*
Neue Zürcher Zeitung, 28., 29., 30. Oktober, (1924)

40 *Über den Ursprung der Empfindlichkeitskurven des Auges*
Die Naturwissenschaften, **12**, (1924), 925-929

41 *Über die Rotationswärme des Wasserstoffs*
Zeitschrift für Physik, **30**, (1924), 341-349

1925

42 *Über Farbenmessung*
Physikalische Zeitschrift, **26**, (1925), 349-352

43 *Die wasserstoffähnlichen Spektren vom Standpunkte der Polarisierbarkeit des Atomrumpfes*
Annalen der Physik, (4), **77**, (1925), 43-70

44 *Über die subjektiven Sternfarben und die Qualität der Dämmerungsempfindung*
Die Naturwissenschaften, **13**, (1925), 373-376

45 *Die Erfüllbarkeit der Relativitätsforderung in der klassischen Mechanik*
Annalen der Physik, (4), **77**, (1925), 325-336

46 *Otto Lummer* †
Neue Zürcher Zeitung, 23. Juli, (1925)

47 *Bemerkungen über die statistische Entropiedefinition beim idealen Gas*
Sitzungsberichte der Preußischen Akademie der Wissenschaften. Physikalisch-mathematische Klasse, (1925), 434-441

47a *Marx, E.: Handbuch der Radiologie 6, Die Theorien der Radiologie* [Book review]
Die Naturwissenschaften, **13**, (1925), 710-711

48 *Michelsonscher Versuch und Relativitätstheorie*
Neue Zürcher Zeitung, 10. September, (1925)

49 *Über das Verhältnis der Vierfarben- zur Dreifarbentheorie*
Sitzungsberichte der Akademie der Wissenschaften in Wien. Mathematisch-naturwissenschaftliche Klasse, Abteilung 2a, **134**, (1925), 471-490

1926

50 *Zur Einsteinschen Gastheorie*
Physikalische Zeitschrift, **27**, (1926), 95-101, Italian: Sulla teoria del gas di Einstein [A 207 j]

51 *Die Energiestufen des idealen einatomigen Gasmodells*
Sitzungsberichte der Preußischen Akademie der Wissenschaften. Physikalisch-mathematische Klasse, (1926), 23-36

52 *Quantisierung als Eigenwertproblem (Erste Mitteilung)*
Annalen der Physik, (4), **79**, (1926), 361-376 [B 1.1]

52a *Guild, J.: Die geometrische Lösung von Farbenmischungsaufgaben* [Article review]
Die Naturwissenschaften, **14**, (1926), 146-147

53 *Quantisierung als Eigenwertproblem (Zweite Mitteilung)*
Annalen der Physik, (4), **79**, (1926), 489-527 [B 1.1]

53a *Herzfeld, K. F.: Kinetische Theorie der Wärme (Müller-Pouillets Lehrbuch der Physik 3/2, 11. Auflage)* [Book review]
Physikalische Zeitschrift, **27**, (1926), 184-185

54 *Das Ehrenfestsche Modell der* H-*Kurve* (with K.W.F. Kohlrausch)
Physikalische Zeitschrift, **27**, (1926), 306-313

55 *Über das Verhältnis der Heisenberg-Born-Jordanschen Quantenmechanik zu der meinen*
Annalen der Physik, (4), **79**, (1926), 734-756 [B 1.1], Hungarian: A Heisenberg-Born-Jordan-féle Kvantummechanika viszonya az enyémhez [A 206a]

56 *Quantisierung als Eigenwertproblem (Dritte Mitteilung: Störungstheorie, mit Anwendung auf den Starkeffekt der Balmerlinien)*
Annalen der Physik, (4), **80**, (1926), 437-490 [B 1.1]

57 *Quantisierung als Eigenwertproblem (Vierte Mitteilung)*
Annalen der Physik, (4), **81**, (1926), 109-139 [B 1.1]

58 *Der stetige Übergang von der Mikro- zur Makromechanik*
Die Naturwissenschaften, **14**, (1926), 664-666 [B 1.1]

59 *An Undulatory Theory of the Mechanics of Atoms and Molecules*
The Physical Review, **28**, (1926), 1049-1070

60 *Spezifische Wärme (theoretischer Teil)*
Handbuch der Physik, **10**, 275-320, Berlin: Springer. 1926.

61 *Die Gesichtsempfindungen*
Müller-Pouillets Lehrbuch der Physik 2/1, 11. Auflage, 456-560, Braunschweig: Vieweg. 1926.

1927

62 *Über den Comptoneffekt*
Annalen der Physik, (4), **82**, (1927), 257-264 [B 1.2]

63 *Der Energieimpulssatz der Materiewellen*
Annalen der Physik, (4), **82**, (1927), 265-272 [B 1.2]

64 *Energieaustausch nach der Wellenmechanik*
Annalen der Physik, (4), **83**, (1927), 956-968 [B 1.2]

1928

65 *Neue Wege in der Physik*
Elektrische Nachrichtentechnik, **5**, (1928), 485-488 [A 68, A 96a]

66 *La mécanique des ondes*
Électrons et Photons. Rapports et Discussions du Cinquième conseil de Physique, 185-213, Paris: Gauthier-Villars. 1928.

1929

67 *Der erkenntnistheoretische Wert physikalischer Modellvorstellungen*
Jahresbericht des Physikalischen Vereins zu Frankfurt am Main 1928/29, (1929), 44-51, English: Conceptual Models in Physics and their Philosophical Value [B 4.1, B 4.2, B 13.1]

68 *Neue Wege in der Physik*
Elektrotechnische Zeitschrift, **50**, (1929), 15-16 [A 65, A 96a]

69 *Was ist ein Naturgesetz?*
Die Naturwissenschaften, **17**, (1929), 9-11 [B 16.1], English: What is a Law of Nature? [B 4.1, B 4.2, B 13.1]

70 *Die Erfassung der Quantengesetze durch kontinuierliche Funktionen*
Die Naturwissenschaften, **17**, (1929), 486-489

71 *Einstein explained*
World's Work, (1929), 52-55; 146

72 *Antrittsrede des Hrn. Schrödinger*
Sitzungsberichte der Preußischen Akademie der Wissenschaften. Physikalisch-mathematische Klasse, (1929), C-CII [A 75, A 96b, A 207b]

73 *Adresse an Hrn. Max Planck zum fünfzigjährigen Doktorjubiläum am 28. Juni 1929* [unsigned]
Sitzungsberichte der Preußischen Akademie der Wissenschaften. Physikalisch-mathematische Klasse, (1929), 341-342

74 *Eddington, A. S.: The Nature of the Physical World* [Book review]
Die Naturwissenschaften, **17**, (1929), 694 [A 94a]

75 *Antrittsrede des Hrn. Schrödinger*
Die Naturwissenschaften, **17**, (1929), 732-733 [abbreviated version of A 72]

76 *Das Rätsel des Lichts*
Die Koralle, **5**, (1929), 294-298

77 *Das Gesetz der Zufälle. Der Kampf um Ursache und Wirkung in den modernen Naturwissenschaften*
Die Koralle, **5**, (1929), 417-418, English: The Law of Chance: The Problem of Causation in Modern Science [B 4.1, B 4.2, B 13.1]

78 *Verwaschene Eigenwertspektra*
Sitzungsberichte der Preußischen Akademie der Wissenschaften. Physikalisch-mathematische Klasse, (1929), 668-682

79 *Vorwort*
Elementare Einführung in die Wellenmechanik; by K. K. Darrow, III-IV, Leipzig: Hirzel. 1929.

1930

79a *Lorentz, H. A.: Vorlesungen über theoretische Physik an der Universität Leiden 4* [Book review]
Deutsche Literaturzeitung, 4. Januar, (1930), 43-44

80 *Das gehetzte Licht*
Uhu (Das neue Ullstein-Magazin), **6**, (1930), 106-112

81 *Was ist eigentlich Elektrizität?*
Die Koralle, **6**, (1930), 110-112

82 *Zum Heisenbergschen Unschärfeprinzip*
Sitzungsberichte der Preußischen Akademie der Wissenschaften. Physikalisch-mathematische Klasse, (1930), 296-303

83 *Über die kräftefreie Bewegung in der relativistischen Quantenmechanik*
Sitzungsberichte der Preußischen Akademie der Wissenschaften. Physikalisch-mathematische Klasse, (1930), 418-428

84 *Wissenschaft-Kunst-Spiel*
Die Koralle, **6**, (1930), 404; 410; 425-426, English: Science as Culture [A 87], English: Science, Art and Play [A 113, B 4.1, B 4.2, B 13.1, A 88]

85 *Verabsäumte Pflichten*
Reclams Universum, **47**, 24. Dezember, (1930), 263

86 *Naturwissenschaft und Ethik*
Vossische Zeitung, 25. Dezember, (1930)

1931

87 *Science as Culture*
The International Forum, 1. Januar, (1931), 10-11 [A 84]

88 *Interviews with great scientists. Prof. Schrödinger*
The Observer, 11. Januar, (1931), 15-16 [A 84]

89 *Zur Quantendynamik des Elektrons*
Sitzungsberichte der Preußischen Akademie der Wissenschaften. Physikalisch-mathematische Klasse, (1931), 63-72

90 *Über die Umkehrung der Naturgesetze*
Sitzungsberichte der Preußischen Akademie der Wissenschaften. Physikalisch-mathematische Klasse, (1931), 144-153

91 *Spezielle Relativitätstheorie und Quantenmechanik*
Sitzungsberichte der Preußischen Akademie der Wissenschaften. Physikalisch-mathematische Klasse, (1931), 238-247

92 *Bemerkung zu der Arbeit des Herrn V. Fock: "Die inneren Freiheitsgrade des Elektrons"*
Zeitschrift für Physik, **70**, (1931), 808-810

92a *Jeffreys, H.: Scientific Inference* [Book review]
Die Naturwissenschaften, **19**, (1931), 967-968

1932

93 *Sur la théorie relativiste de l'électron et l'interprétation de la mécanique quantique*
Annales de l'Institut Henri Poincaré, **2**, (1932), 269-310

94 *Anmerkungen zum Kausalproblem*
Erkenntnis (zugleich Annalen der Philosophie), **3**, (1932), 65-70

94a *Eddington, A. S.: Das Weltbild der Physik und ein Versuch seiner philosophischen Deutung* [Book review]
Die Naturwissenschaften, **20**, (1932), 172-173 [A 74, with a comment of the editor]

95 *Diracsches Elektron im Schwerefeld I*
Sitzungsberichte der Preußischen Akademie der Wissenschaften. Physikalisch-mathematische Klasse, (1932), 105-128

95a *Sommerfeld, A.: Atombau und Spektrallinien, 5. Auflage* (with V. F. Weisskopf) [Book review]
Die Naturwissenschaften, **20**, (1932), 332

96 *Über das Verhalten des Starkeffekts bei plötzlichen Feldänderungen* (Experimenteller Teil von H. Rausch von Traubenberg und R. Gebauer; theoretischer Teil von E. Schrödinger)
Zeitschrift für Physik, **78**, (1932), 309-317

96a *Neue Wege in der Physik*
Materie und Energie vom naturwissenschaftlichen Weltbild der Gegenwart; edited and commentated by E. Wildhagen, 348-354. Berlin: Deutsche Buchgemeinschaft. 1932. [A 65, A 68]

96b [*Antrittsrede des Hrn. Schrödinger*]
Materie und Energie vom naturwissenschaftlichen Weltbild der Gegenwart; edited and commentated by E. Wildhagen, 355-362. Berlin: Deutsche Buchgemeinschaft. 1932. [A 72, A 207b]

1933

97 *L'électron de Dirac dans la theorie de la relativité générale*
Comptes Rendus du Congrès International d'Électricité Paris 1932 (Première Section), 581-591, Paris: Gauthier-Villars. 1933.

97a *van der Waerden, B. L.: Die gruppentheoretische Methode in der Quantenmechanik* [Book review]
Physikalische Zeitschrift, **34**, (1933), 184

98 *Über den zweiten Hauptsatz der Thermodynamik*
Sitzungsberichte der Preußischen Akademie der Wissenschaften. Physikalisch-mathematische Klasse, (1933), 165

99 *Warum sind die Atome so klein?*
Forschungen und Fortschritte, **9**, (1933), 125-126

1934

99a *Lorentz, H. A.: Vorlesungen über theoretische Physik an der Universität Leiden 5* [Book review]
Deutsche Literaturzeitung, 17. Juni, (1934), 1146

100 *Über die Unanwendbarkeit der Geometrie im Kleinen*
Die Naturwissenschaften, **22**, (1934), 518-520

101 *Der Grundgedanke der Wellenmechanik*
Die moderne Atomtheorie; die bei der Entgegennahme des Nobelpreises 1933 in Stockholm gehaltenen Vorträge, 19-36, Leipzig: Hirzel. 1934. [A 110, B 16.1], English: The Fundamental Idea of Wave Mechanics [B 4.1, B 4.2., B 13.1], Japanese: in: Nôberu-shô kôen. Butsurigaku 5, Tôkyô. Kôdansha. 1978.

1935

102 *The Absolute Field Constant in the New Field Theory* (with M. Born)
Nature, **135**, (1935), 342

103 *Contributions to Born's New Theory of the Electromagnetic Field*
Proceedings of the Royal Society of London, A, **150**, (1935), 465-477

104 *Quelques remarques au sujet des bases de la connaissance scientifique*
Scientia (Rivista di Scienza), **57**, (1935), 181-191

105 *Equality and Relativity of Freedom*
The Listener, 5. June, (1935), 952-953, German: Gleichheit und Relativität der Freiheit [A 106b]

106 *¿Son lineales las verdaderas ecuaciones del campo electromagnético?*
Anales de la Sociedad Española de Fisica y Quimica, **33**, (1935), 511-517

106a *Frank, Ph. and von Mises, R.: Die Differential- und Integralgleichungen der Mechanik und Physik 2, Physikalischer Teil* [Book review]
Die Naturwissenschaften, **23**, (1935), 516-517

106b *Gleichheit und Relativität der Freiheit*
Die Auslese aus Zeitschriften des In- und Auslandes, **9**, (1935), 650-652, [abbreviated translation of A 105]

107 *Discussion of probability relations between separated systems*
Proceedings of the Cambridge Philosophical Society, **31**, (1935), 555-563

108 *Die gegenwärtige Situation in der Quantenmechanik*
Die Naturwissenschaften, **23**, (1935), 807-812; 823-828; 844-849 [A 207d]

109 *La nueva mecánica ondulatoria*
'Cursos de la Universidad Internacional de Verano en Santander, **1**, 1-73, Madrid: Signo. 1935.

109a *The Decline*
Time and Tide (University Supplement), 30. November, (1935), 1757-1758

110 *Der Grundgedanke der Wellenmechanik*
Les Prix Nobel en 1933, 1-13, Stockholm: Norstedt & Söner. 1935. [A 101]

111 [*Trinkspruch*]
Les Prix Nobel en 1933, 79-81, Stockholm: Norstedt & Söner. 1935.

112 *Erwin Schrödinger* [autobiographical sketch]
Les Prix Nobel en 1933, 86-88, Stockholm: Norstedt & Söner. 1935.

113 *Science, Art and Play*
The Philosopher (London), **13**, (1935), 11-18 [A 84]

1936

114 *Probability relations between separated systems*
Proceedings of the Cambridge Philosophical Society, **32**, (1936), 446-452

115 *Phenomenological Theory of Supra-conductivity*
Nature, **137**, (1936), 824

116 *Indeterminism and Free Will*
Nature, **138**, (1936), 13-14

1937

117 *World Structure*
Nature, **140**, (1937), 742-744

117a *Über die Eddington'sche Welttheorie - Sulla teoria dell'universo di Eddington.* [Extract]
Celebrazione del secondo centenario della nascita di Luigi Galvani. Congresso della Societa Italiana di Fisica, 17-20 Octobre 1937, A. XV, N. 12, Bologna: Societa Italiana di Fisica. 1937. [A 119]

1938

118 *Eigenschwingungen des sphärischen Raumes*
Commentationes Pontificiae Academiae Scientiarum, **2**, (1938), 321-364

119 *Sur la théorie du monde d'Eddington*
Il Nuovo Cimento, (N.S.), **15**, (1938), 246-254 [A 117a]

120 *Die Mehrdeutigkeit der Wellenfunktion*
Annalen der Physik, (5), **32**, (1938), 49-55

121 *Mean Free Path of Protons in the Universe*
Nature, **141**, (1938), 410

1939

122 *Nature of the Nebular Red-Shift*
Nature, **144**, (1939), 593

123 *The proper vibrations of the expanding universe*
Physica, **6**, (1939), 899-912

1940

124 *A Method of Determining Quantum-Mechanical Eigenvalues and Eigenfunctions*
Proceedings of the Royal Irish Academy, **46 A**, (1940), 9-16

124a *Eddington, A. S.: The Philosophy of Physical Science (Tarner Lectures, 1938)* [Book review]
Nature, **145**, (1940), 402-403

125 *Maxwell's and Dirac's Equations in the Expanding Universe*
Proceedings of the Royal Irish Academy, **46 A**, (1940), 25-47

126 *Boolean Algebra and Probability Theory* (with T. S. Broderick)
Proceedings of the Royal Irish Academy, **46 A**, (1940), 103-112

127 *The General Theory of Relativity and Wave Mechanics*
Wis-en natuurkundig Tijdschrift, **10**, (1940), 2-9 [A 183]

1941

128 *Prof. Richard Bär* [Obituary]
Nature, **147**, (1941), 536

129 *Further Studies on Solving Eigenvalue Problems by Factorization*
Proceedings of the Royal Irish Academy, **46 A**, (1941), 183-206

130 *On the Solutions of Wave Equations for Non-Vanishing Rest-Mass Including a Source-Function*
Proceedings of the Royal Irish Academy, **47 A**, (1941), 1-23

131 *Exchange and Spin* (with a Note by James Hamilton)
Proceedings of the Royal Irish Academy, **47 A**, (1941), 39-52

132 *The Factorization of the Hypergeometric Equation*
Proceedings of the Royal Irish Academy, **47 A**, (1941), 53-54

133 *La structure de l'Univers en relation avec la structure corpusculaire*
Bulletin de la Société Philomathique de Paris, **123**, (1941), 26-30

1942

134 *Non-linear Optics*
Proceedings of the Royal Irish Academy, **47 A**, (1942), 77-117

135 *Dynamics and Scattering-power of Born's Electron*
Proceedings of the Royal Irish Academy, **48 A**, (1942), 91-122

1943

136 *Pentads, Tetrads, and Triads of Meson-Matrices*
Proceedings of the Royal Irish Academy, **48 A**, (1943), 135-146

137 *Systematics of Meson-Matrices*
Proceedings of the Royal Irish Academy, **49 A**, (1943), 29-42

138 The *General Unitary Theory of the Physical Fields*
Proceedings of the Royal Irish Academy, **49 A**, (1943), 43-58

139 *A new Exact Solution in Non-Linear Optics (Two-Wave-System)*
Proceedings of the Royal Irish Academy, **49 A**, (1943), 59-66

140 *The Earth's and the Sun's Permanent Magnetic Fields in the Unitary Field Theory*
Proceedings of the Royal Irish Academy, **49 A**, (1943), 135-148

1944

141 *The Point Charge in the Unitary Field Theory*
Proceedings of the Royal Irish Academy, **49 A**, (1944), 225-235

142 *Unitary Field Theory: Conservation Identities and Relation to Weyl and Eddington*
Proceedings of the Royal Irish Academy, **49 A**, (1944), 237-244

143 *The Shielding Effect of Planetary Magnetic Fields* (with J. McConnell)
Proceedings of the Royal Irish Academy, **49 A**, (1944), 259-273

144 *The Union of the three Fundamental Fields (Gravitation, Meson, Electromagnetism)*
Proceedings of the Royal Irish Academy, **49 A**, (1944), 275-287

145 *The Affine Connexion in Physical Field Theories*
Nature, **153**, (1944), 572-575

146 *Rate of n-fold Accidental Coincidences*
Nature, **153**, (1944), 592-593

147 *The Statistical Law in Nature*
Nature, **153**, (1944), 704-705

1945

148 *On Distant Affine Connection*
Proceedings of the Royal Irish Academy, **50 A**, (1945), 143-154

149 *Infinitesimal Affine Connections with Twofold Einstein-Bargmann Symmetry* (with F. Mautner)
Proceedings of the Royal Irish Academy, **50 A**, (1945), 223-231

150 *Probability Problems in Nuclear Chemistry*
Proceedings of the Royal Irish Academy, **51 A**, (1945), 1-8

151 *Österreichische Wissenschaft*
Kulturelle Schriftenreihe des FAM (Free Austrian Movement), London, (1945), 1-3 [A 152], English: Austrian Science [A 152a]

152 *Österreichische Wissenschaft*
Austro American Tribune. Anti-Nazi Monthly, **4**, (1945), 7 [slightly modified version of A 151]

152a *Austrian Science*
Science in Austria. Leaflet, presented on the occasion of the meeting of British and Austrian scientists in support of the restoration of Science in Austria. 12-13, London: Association of Austrian Engineers, Chemists and Scientific Workers in Great Britain. 1945. [A 151]

1946

153 *The General Affine Field Laws*
Proceedings of the Royal Irish Academy, **51 A**, (1946), 41-50

154 *Affine Feldtheorie und Meson*
Verhandlungen der Schweizerischen Naturforschenden Gesellschaft, **126**, (1946), 53-61

1947

155 *The Foundation of the Theory of Probability - I*
Proceedings of the Royal Irish Academy, **51 A**, (1947), 51-66

156 *The Foundation of the Theory of Probability - II*
Proceedings of the Royal Irish Academy, **51 A**, (1947), 141-146

157 *The relation between Metric and Affinity*
Proceedings of the Royal Irish Academy, **51 A**, (1947), 147-150

158 *The Final Affine Field Laws I*
Proceedings of the Royal Irish Academy, **51 A**, (1947), 163-171

159 *Der Geist der Naturwissenschaft*
Eranos-Jahrbuch 1946, **14**, 491-520, Zürich: Rhein. 1947. [A 207], English: The Spirit of Science [A 189, B 12]

1948

160 *2400 Jahre Quantentheorie*
Annalen der Physik, (6), **3**, (1948), 43-48, Hungarian: A 2400 éves kvantumelmélet [A 204]

161 *Die Besonderheit des Weltbilds der Naturwissenschaft*
Acta Physica Austriaca, **1**, (1948), 201-245 [B 16.1], English: On the Peculiarity of the Scientific World-View [B 12]

162 *The Final Affine Field Laws II*
Proceedings of the Royal Irish Academy, **51 A**, (1948), 205-216

163 *The Final Affine Field Laws III*
Proceedings of the Royal Irish Academy, **52 A**, (1948), 1-9

164 *Theoretiker und Praktiker*
Die Furche, 27. März, (1948)

1950

165 *What is an elementary particle?*
Endeavour, **9**, (1950), 109-116 [A 174, B 13.1], German: Was ist ein Elementarteilchen? Endeavour, **9**, (1950), 109-118 [A 173, B 16.1], French: Qu'est-ce qu'une particule élémentaire? Endeavour, **9**, (1950), 109-116, Spain: La particula elemental. Endeavour, **9**, (1950), 109-116, Italian: Cos'è una particella elementare? Endeavour, **9**, (1950), 109-116, Polish: Co to jest czastka elementarna? [A 203]

166 *Irreversibility*
Proceedings of the Royal Irish Academy, **53 A**, (1950), 189-195 [A 207c]

167 *The Future of Understanding - Die Zukunft des Weltverstehens*
Three BBC Talks on September 16, 23, 30, (1950) [B 12, B 14.1, B 14.3]

1951

168 *Studies in the Non-Symmetric Generalization of the Theory of Gravitation I*
Communications of the Dublin Institute for Advanced Studies, Series A, **6**, (1951), 28 S.

169 *On the Differential Identities of an Affinity*
Proceedings of the Royal Irish Academy, **54 A**, (1951), 79-85

170 *The Point-Charge in the Non-symmetric Field Theory* (with A. Papapetrou)
Nature, **168**, (1951), 40-41

171 *Studies in the Generalized Theory of Gravitation II: The Velocity of Light* (with O. Hittmair)
Communications of the Dublin Institute for Advanced Studies, Series A, **8**, (1951), 15 S.

172 *A Combinatorial Problem in Counting Cosmic Rays*
The Proceedings of the Physical Society, Section A, **64**, (1951), 1040-1041

173 *Was ist ein Elementarteilchen?*
Die Pyramide, (1951), 2-4; 24-25; 44-46 [A 165]

174 *What is an elementary particle?*
The Smithsonian Institution's Annual Report, 183-196, Washington: U.S. Government Printing Office. 1951. [A 165]

1952

175 *Dirac's New Electrodynamics*
Nature, **169**, (1952), 538

176 *Are There Quantum Jumps? Part I*
The British Journal for the Philosophy of Science, **3**, (1952), 109-123 [B 12]

177 *Are There Quantum Jumps? Part II*
The British Journal for the Philosophy of Science, **3**, (1952), 233-242 [B 12]

178 *Relativistic Fourier Reciprocity and the Elementary Masses*
Proceedings of the Royal Irish Academy, **55 A**, (1952), 29-50

1953

179 *L'image actuelle de la matière (Sommaire) - Unsere Vorstellung von der Materie*
L'homme devant la science, Texte des conférences et des entretiens organisés par les rencontres internationales de Genève 1952, 31-54, Neuchâtel: Baconnière. 1953. German: Unsere Vorstellung von der Materie [A 180, A 187, B 16.1], English: Our Conception of Matter [B 12], What is Matter? [A 181, A 208], Our Image of Matter [A 205], Italian: L'immagine attuale della materia [A 202]

180 *Unsere Vorstellung von der Materie*
Merkur, **7**, (1953), 131-145 [A 179, with a preamble by the editors, A 207e]

181 *What is Matter?*
Scientific American, **189**, (1953), 52-57 [abbreviated version of "Our Conception of Matter" A 179, A 208]

182 *The Meaning of Wave Mechanics - La signification de la mécanique ondulatoire*
Louis de Broglie, Physicien et Penseur, 16-32, Paris: Michel. 1953. German: Die Bedeutung der Wellenmechanik. Louis de Broglie und die Physiker, 18-25, Hamburg: Claassen. 1955.

183 *The General Theory of Relativity and Wave Mechanics*
Scientific Papers Presented to Max Born, 65-74, Edinburgh: Oliver & Boyd. 1953. [A 127]

1954

184 *Electric Charge and Current engendered by combined Maxwell-Einstein-Fields*
Proceedings of the Royal Irish Academy, **56 A**, (1954), 13-21

185 *Relativistic Quantum Theory*
The British Journal for the Philosophy of Science, **4**, (1954), 328-329 [Extract from a private letter]

186 *Measurement of Length and Angle in Quantum Mechanics*
Nature, **173**, (1954), 442

186a *Begegnung; Parabel*
Kontinente. Wege und Probleme der Gegenwart, **7**, (1954), 28
[two poems of B 7.1, A 207l, A 207m, A 207n, A 207p]

187 *Unsere Vorstellung von der Materie*
Naturwissenschaftliche Rundschau, **7**, (1954), 277-282, [abbreviated version of A 179]

188 *Orientierung im Weltall; Erdalter und Weltalter; Die Kohlenstoff-Uhr; Raum und Zeit*
Orientierung im Weltall, 7-31, Zürich: Fontana. 1954. (Das Internationale Forum. Berichte und Stellungnahmen. 3)

189 *The Spirit of Science*
Spirit and Nature, Papers from the Eranos Yearbooks, 322-341, New York: Pantheon Books. 1954. [A 159]

1955

190 *The Philosophy of Experiment*
Il Nuovo Cimento, (10), **1**, (1955), 5-15

191 *A Thermodynamic Relation between Frequency-Shift and Broadening*
Il Nuovo Cimento, (10), **1**, (1955), 63-69

192 *The wave equation for spin 1 in Hamiltonian form* [*I*]
Proceedings of the Royal Society of London, A, **229**, (1955), 39-43

193 *Atomenergie*
Sie und er, 27. Januar, (1955), 20-22

194 *Must the photon mass be zero?* (with L. Bass)
Proceedings of the Royal Society of London, A, **232**, (1955), 1-6 [A 197]

195 *The wave equation for spin 1 in Hamiltonian form. II*
Proceedings of the Royal Society of London, A, **232**, (1955), 435-447

196 *Die Atomisten*
Merkur, **9**, (1955), 815-824 [B 10.2], English: The atomists [B 10.1]

1956

197 *Must the Photon Mass be Zero?* (with L. Bass), [Summary and discussion]
Il Nuovo Cimento, Supplemento, (10), **4**, (1956), 825-826 [A 194]

1957

198 [*Festrede, gehalten bei der Eröffnung der fünften Weltkraftkonferenz, Wien 1956*]
Fünfte Weltkraftkonferenz, Wien 1956, Gesamtbericht, Band I, 277-283 [German], 283-289 [English], 289-295 [French], Wien: Österreichisches Nationalkomitee der Weltkraftkonferenz. 1957. [B 14.1, B 14.3, B 15.1, B 15.2]

199 *Zur Geistesgeschichte der Stellung der Menschen*
Der Mittelschullehrer und die Mittelschule, **6**, (1957), 280-282

200 *Die Atomtheorie*
Lebendige Stadt. Almanach 1957, 157-161, Wien: Amt für Kultur und Volksbildung der Stadt Wien. 1957.

1958

201 *Might perhaps Energy be a merely Statistical Concept?*
Il Nuovo Cimento, (10), **9**, (1958), 162-170

1959

201a *Commilitonen! Upanishad*
Quo via fert? Wohin führt der Weg, 10, Wien: Maturanten des Döblinger Gymnasiums, Wien XIX. 1959. [Short letter and unpublished poem]

202 *L'immagine attuale della materia*
Discussione sulla Fisica Moderna, 35-57, Torino: Boringhieri. 1959. 1960. 1964 and 1980. [A 179]

1960

203 *Co to jest czastka elementarna?*
Postepy Fizyki, **11**, (1960), 135-150 [A 165]

1961

203a *Religion und Naturwissenschaft*
Physikalische Blätter, **17**, (1961), 105-110 [Short part of B 10.2]

204 *A 2400 éves kvantumelmélet*
Fizikai Szemle, **11**, (1961), 101-104 [A 160]

205 *Our Image of Matter*
On Modern Physics, 45-66, New York: Clarkson N. Potter. 1961. London: Orion Press. 1961. New York: Crowell-Collier Publishing Company. 1962. [A 179]

1962

206 *Die Wandlung des physikalischen Weltbegriffs*
[Lecture delivered at Deutsches Museum, Munich, 6th May 1930; B 16.1]

1966

206a *A Heisenberg - Born - Jordan-féle kvantummechanika viszonya az enyémhez*
Magyar Fizikai Folyóirat, **14**, (1966), 359-374 [A 55]

207 *Der Geist der Naturwissenschaft*
Gibt es Grenzen der Naturforschung? 15-36, Freiburg, Basel, Wien: Herder. 1966. (Herder-Bücherei 253) [A 159]

1975

207a *Fragment from an unpublished dialogue of Galileo*
Hermathena. A Dublin University Review, **119**, (1975), 74-77
[originally published in the King's Hospital School Magazine]

1979

207b *Antrittsrede von Erwin Schrödinger (1887-1961). Erwiderung von Max Planck*
Physiker über Physiker 2. Antrittsreden, Erwiderungen bei der Aufnahme von Physikern in die Berliner Akademie. Gedächtnisreden. 1870-1929; adapted by Ch. Kirsten and H.-G. Körber, 264-268, Berlin: Akademie. 1979. (Studien zur Geschichte der Akademie der Wissenschaften der DDR 8) [A 72, A 96b]

1982

207c *Irreversibility*
The Enigma of Time; compiled by P. T. Landsberg, 46-52, Bristol, Boston: Hilger. 1982, 1984 and 1985. [A 166]

1984

207d *Die gegenwärtige Situation in der Quantenmechanik (1935)*
Die Deutungen der Quantentheorie; by K. Baumann and R. U. Sexl, 98-129, Braunschweig: Vieweg. 1984. (Facetten der Physik 11) [A 108]

1986

207e *Was ist Materie?*
Elementare Materie, Vakuum und Felder; edited by W. Greiner and G. Wolschin, 28-35, Heidelberg: Spektrum der Wissenschaft. 1986. (Spektrum der Wissenschaft: Verständliche Forschung) [slightly abridged version of A 180]

207f *Das arithmetische Paradoxon - Die Einheit des Bewußtseins*
Physik und Transzendenz. Die großen Physiker unseres Jahrhunderts über ihre Begegnung mit dem Wunderbaren; edited by H.-P. Dürr, 159-170, Bern, München, Wien: Scherz. 1986. [B 14.3]

207g *Naturwissenschaft und Religion*
Physik und Transzendenz. Die großen Physiker unseres Jahrhunderts über ihre Begegnung mit dem Wunderbaren; edited by H.-P. Dürr, 171-183, Bern, München, Wien: Scherz. 1986. [B 14.3]

207h *Was ist wirklich? - Die Gründe für das Aufgeben des Dualismus von Denken und Sein oder von Geist und Materie*
Physik und Transzendenz. Die großen Physiker unseres Jahrhunderts über ihre Begegnung mit dem Wunderbaren; edited by H.-P. Dürr, 184-188, Bern, München, Wien: Scherz. 1986. [B 15.1, B 22.1]

207i *Die vedantische Grundansicht*
Physik und Transzendenz. Die großen Physiker unseres Jahrhunderts über ihre Begegnung mit dem Wunderbaren; edited by H.-P. Dürr, 189-192, Bern, München, Wien: Scherz. 1986. [B 15.1, B 22.1]

207j *Sulla teoria del gas di Einstein*
La statistica quantistica e le onde di materia; edited by P. Bernardini, Napoli: Bibliopolis. 1986. [A 50]

1987

207k *Beruht Leben auf physikalischen Gesetzen?*
Leben = Physik + Chemie? Das Lebendige aus der Sicht bedeutender Physiker; edited by B.-O. Küppers, 73-83, München, Zürich: Piper. 1987. (Serie Piper 599) [B 5.2, B 5.3, B 5.3a]

2071 *Parabel*
In memoriam. Lajos Jánossy - 75. Erwin Schrödinger - 100; edited by P. Király and M. Ziegler-Náraynè, XIII, Budapest: MTA Központi Fizikai Kutató Intézete. 1987. [poem of B 7.1, A 186a, A 207m, A 207n, A 207p], Hungarian: Parabola [translation in the same publication on page XII]

207m *Parabel*
Jánossy Lajos (1912-1978) és Erwin Schrödinger (1887-1961) levelezése; edited by P. Király and M. Ziegler-Náraynè, IX, Budapest: MTA Központi Fizikai Kutató Intézete. 1987. [poem of B 7.1, A 186a, A 2071, A 207n, A 207p], Hungarian: Parabola [translation in the same publication on page VIII]

207n *Parabel*
Physikalische Blätter, **43**, (1987), 335
[poem of B 7.1, A 186a, A 2071, A 207m, A 207p]

1989

207o *Zur Verteidigung der universalen Betrachtungsweise*
Lust am Forschen. Ein Lesebuch zu den Naturwissenschaften; edited by K. Stadler, 17, München, Zürich: Piper. 1989. (Serie Piper 1050) [Preface to B 5.3, B 5.3a]

207p *Parabel; Zürich; Liebeslied; Geborgen; Der Entäuschte; On the shore; Juni*
Schrödinger, life and thought; by W. Moore, 6, 151, 408, 410, 413, 419-420, 447, Cambridge: University Press. 1989. [Poems of B 7.1 together with a translation into English]

207q *Zittern*
Schrödinger, life and thought; by W. Moore, 422, Cambridge: University Press. 1989. [unpublished poem]

207r *Herbst*
Schrödinger, life and thought; by W. Moore, 444, Cambridge: University Press. 1989. [unpublished poem]

207s *Glückliche Kindheit*
SOS-Kinderdorf-Jahrbuch, 12-13, Innsbruck: SOS-Kinderdorf. 1989. [Short part of B 22.1]

[Without reference to the year of publication]

208 *What Is Matter?*
Supplementary Readings for Chemical Bond Approach, 2-8 [24 articles reprinted from Scientific American, A 179, A 181]

209 *Infinites - A Discourse on Transfinite Numbers*
The Times Review of the Progress of Science [no more bibliographic data available]

B. BOOKS

1927

1.1 *Abhandlungen zur Wellenmechanik*
Leipzig: Barth. 1927.
Vorwort und sachlich geordnete Inhaltsangabe, Seite III, V - X
Quantisierung als Eigenwertproblem (1. Mitteilung), 1-16 [A 52]
Quantisierung als Eigenwertproblem (2. Mitteilung), 17-55 [A 53]
Der stetige Übergang von der Mikro- zur Makromechanik, 56-61 [A 58]
Über das Verhältnis der Heisenberg-Born-Jordanschen Quantenmechanik zu der meinen, 62-84 [A 55]
Quantisierung als Eigenwertproblem (3. Mitteilung), 85-138 [A 56]
Quantisierung als Eigenwertproblem (4. Mitteilung), 139-169 [A 57]

1.2 *Abhandlungen zur Wellenmechanik*
Leipzig: Barth. 1928. Zweite, vermehrte Auflage.
[To the first edition the following three articles were added]
Über den Comptoneffekt, 170-177 [A 62]
Der Energieimpulssatz der Materiewellen, 178-185 [A 63]
Energieaustausch nach der Wellenmechanik, 186-198 [A 64]

1.3 *Die Wellenmechanik*
Stuttgart: Battenberg. 1963. (Dokumente der Naturwissenschaften, Abteilung Physik, Band 3, edited by A. Hermann)
[The book contains: A 52, A 53, A 55, A 56, A 57]

1.4 *Hadô-rikigaku ronbunshû*
[Translation of *Die Wellenmechanik* into Japanese by Tanaka Shô and Minami Masatsugu]
Tôkyô: Kyôritsu shuppan. 1974. (Schrödinger senshû)

1.5 *Wellenmechanik*. Einführung und Originaltexte. Von G. Ludwig
[Translation of the volume "Wave Mechanics" from the series "Selected Readings in Physics" by D. ter Haar]
Berlin: Akademie-Verlag. Oxford: Pergamon-Press. 1968. Braunschweig: Vieweg. 1969. (WTB-Wissenschaftliche Taschenbücher, Band 55)
[The book contains: A 52, A 53, A 55, A 57]

1.6 *Collected Papers on Wave Mechanics*
[Translation of the second german edition [B 1.2] into English by J. F. Shearer and W. M. Deans]
London: Blackie and Son. 1928.
Reprint: London: Blackie and Son. 1929.
New York: Chelsea Publishing Company. 1978. [B 21]

1.7 *Mémoires sur la mécanique ondulatoire*
[Translation of the second german edition [B 1.2] into French by A. Proca, "Préface" by M. Brillouin, *Avant-propos* and *Additions* by E. Schrödinger]
Paris: Alcan. 1933.
Reprint: Paris: Jacques Gabay. 1988.

1.8 *La Nueva Mecánica ondulatoria*
[Translation into Spanish by X. Zubiri]
Madrid: Signo. 1935. (Cursos de la Universidad internacional de verano en Santander 1)

1928

2.1 *Four Lectures on Wave Mechanics*, delivered at the Royal Institution, London on 5th, 7th, 12th and 14th March, 1928
London, Glasgow: Blackie and Son. [1928].
Reprint: London, Glasgow: Blackie and Son. 1929. [B 21]
First Lecture, 1-13
Second Lecture, 14-26
Third Lecture, 27-42
Fourth Lecture, 43-53

2.2 *Vier Vorlesungen über Wellenmechanik*
[Translation into German by H. Kopfermann]
Berlin: Springer. 1928.

2.3 *Cetyre lekcii po volnovoj mechanike*
[Translation into Russian]
Char'kow-Kiew: 1936.

1932

3.1 *Über Indeterminismus in der Physik - Ist die Naturwissenschaft milieubedingt?* Zwei Vorträge zur Kritik der naturwissenschaftlichen Erkenntnis
Leipzig: Barth. 1932.

3.2 *Zagadnienia współczesnej nauki - Indeterminizm. Wpływ srodowiska na nauki przyrodnicze*
[Translation into Polish by E. Poznanski]
Warszawa: Mathesis Polskiej. 1933.
[The small volume also contains the article by M. Planck "Der Kausalbegriff in der Physik" translated into Polish]

3.3 [The translation of *Über Indeterminismus in der Physik* into English - *Indeterminism in Physics* - has been done by W. H. Johnston [B 4.1, B 4.2, B 13.1]; Schrödinger's text *Ist die Naturwissenschaft milieubedingt?* was rewritten in English by J. Murphy for the essays *Is Science a Fashion of the Times?* and *Physical Science and the Temper of the Age.* [B 4.1, B 4.2, B 13.1]]

1935

4.1 *Science and the Human Temperament* (translated and with a biographical introduction by James Murphy; a foreword by Lord Rutherford of Nelson)
London: Allen and Unwin. 1935. [B 13.1]
Science, Art and Play, 23-32 [A 84]
The Law of Chance: The Problem of Causation in Modern Science, 33-42 [A 77]
Indeterminism in Physics, 43-65 [B 3.3, B 3.1]
Is Science a Fashion of the Times? 66-85 [B 3.3, B 3.1]
Physical Science and the Temper of the Age, 86-106 [B 3.3, B 3.1]
What is a Law of Nature? 107-118 [A 69]
Conceptual Models in Physics and their Philosophical Value, 119-132 [A 67]
The Fundamental Idea of Wave Mechanics, 133-154 [A 101]

4.2 *Science and the Human Temperament* (translated by Dr. James Murphy and W. H. Johnston)
New York: Norton. 1935.

1944

5.1 *What is Life*? The Physical Aspect of the Living Cell. Based on Lectures delivered under the auspices of the Institute at Trinity College, Dublin, in February 1943
Cambridge: University Press. 1944. [B 12, B 18a.1]
Reprint: Cambridge: University Press. 1945. 1948. 1951. 1955. 1962 and 1967.
New York: Macmillan. 1945.
Reprint: New York: Macmillan. 1946. 1947.
The Classical Physicist's Approach to the Subject, 1-17
The Hereditary Mechanism, 18-31
Mutations, 32 45
The Quantum-Mechanical Evidence, 46-55
Delbrück's model discussed and tested, 56-67
Order, Disorder and Entropy, 68-75
Is Life based on the Laws of Physics? 76-87
Epilogue. On Determinism and Free Will, 88

5.2 *Was ist Leben?* Die lebende Zelle mit den Augen des Physikers betrachtet
[Translation into German by L. Mazurczak]
Bern: Francke. 1946. (Sammlung Dalp 1)
München: Lehnen. 1946.

5.2a *Che cos'è la vita?*
[Translation into Italian by M. Ageno]
Firenze: Sansoni. 1947.

5.3 *Was ist Leben?* Die lebende Zelle mit den Augen des Physikers betrachtet
[Second edition: Translation into German by L. Mazurczak and revised by E. Schneider; according to E. Schrödinger it is the only useful German edition available]
Bern: Francke. 1951.
München: Lehnen. 1951. (Sammlung Dalp 1)

5.3a *Was ist Leben?* Die lebende Zelle mit den Augen des Physikers betrachtet
[New edition of B 5.3 with corrections and an introduction "'Was ist Leben?' - mehr als vierzig Jahre später" by E. P. Fischer]
München, Zürich: Piper. 1987.
München, Zürich: Piper. 1989. (Serie Piper 1134)

5.4 *¿Qué es la vida?*
[Translation into Spanish by G. Mayena]
Buenos Aires: Espasa - Calpe. 1948.
[Translation into Spanish by R. Guerrero]
Barcelona: Avance. 1976. (Avance Teoría 6)
Barcelona: Tusquets. 1983. (Cuadernos infimos 107)

5.5 *Qu'est-ce que la vie?* L'aspect physique de la cellule vivante
[Translation into French by L. Keffler]
Paris: Club français du livre. 1949.
Bruxelles: Editions de la Paix. 1951. (Actualité scientifique)
Paris: Bourgois. 1986. (Épistémè)

5.6 *Vad är liv? Den levande cellen ur fysikalisk synpunkt*
[Translation into Swedish by E. R. Ygberg]
Stockholm: Bonnier. 1949.

5.7 *Seimei towa nanika*
[Translation into Japanese by Shôten Oka and Yasuo Shizume]
Tôkyô: Iwanami shoten. 1951.

5.8 *Che cos'è la vita?*
[with Scienza e umanesimo [B 9.4] added: Translation into Italian by P. Lantermo und M. Ageno]
Firenze: Sansoni. 1970.

5.9 *Cto takoe žizn'?*
[Translation into Russian by A. A. Malinovskij and G. G. Porošenko]
Moskva: Atomizdat. 1972. [2. Russian edition]
[Already in 1947 a Russian translation of *What is Life?* was published in Moscow.]

6.1 *Statistical Thermodynamics.* A Course of Seminar Lectures delivered in January-March 1944, at the School of Theoretical Physics, Dublin Institute for Advanced Studies
Dublin: Institute for Advanced Studies. 1944. (Hectograph)
Cambridge: University Press. 1946.
Toronto: Macmillan. 1946.
Reprint: Cambridge: University Press. 1948.
General introduction, 1-4
The method of the most probable distribution, 5-14
Discussion of the Nernst theorem, 15-17
Examples on the second section, 18-21
Fluctuations, 22-26
The method of mean values, 27-41
The n-particle problem, 42-52
Evaluation of the formulae. Limiting cases, 53-80
The problem of radiation, 81-88

6.1a *Statistical Thermodynamics.* A Course of Seminar Lectures delivered in January-March 1944, at the School of Theoretical Physics, Dublin Institute for Advanced Studies
[Second Edition: Except for the *Appendix* the Second Edition is a reprint of B 6.1]
Cambridge: University Press. 1952.
Reprint: Cambridge: University Press. 1957. 1960.
New York: Dover. 1989.

6.2 *Statistische Thermodynamik*
[Translation into German by W. Bloch]
Leipzig: Barth. 1952.
Braunschweig: Vieweg. 1978.

6.3 *Statisticeskaja termodinamica*
[Translation into Russian]
Moskau: 1948.

6.4 *Termodinamica statistica*
[Translation into Italian]
Torino: Boringhieri. 1961.

1949

7.1 *Gedichte*
Godesberg: Küpper. 1949.

7.2 *Poesie*
[together with *La mia visione del mondo. La mia vita.* [B 22.3] All poems were reprinted and the larger part of them translated by either H. Fachinelli or E. V. de Regny into Italian]
Milano: Garzanti. 1987. (Saggi rossi)

1950

8.1 *Space-Time Structure*
Cambridge: University Press. 1950.
Reprint with corrections: Cambridge: University Press. 1954. 1960. 1985. 1986.
Introduction, 1-3
The Unconnected Manifold, 4-26
Affinely Connected Manifold, 27-62
Metrically Connected Manifold, 63-119

8.2 *Jikû no kôzô*
[Translation into Japanese by Uchiyama Ryôyu and Takabayashi Takehiko]
Tôkyô: Kyôritsu shuppan. 1974. (Schrödinger senshû 2)

8.3 *Die Struktur der Raum-Zeit*
[Translation into German and epilogue by J. Audretsch]
Darmstadt: Wissenschaftliche Buchgesellschaft. 1987.

1951

9.1 *Science and Humanism*. Physics in Our Time
Cambridge: University Press. 1951.
Reprint: Cambridge: University Press. 1952.
Preface, ix
The spiritual bearing of science on life, 1-9 [B 12]
The practical achievements of science tending to obliterate its true import, 9-11 [B 12]
A radical change in our ideas of matter, 11-18
Form, not substance, the fundamental concept, 18-21
The nature of our 'models', 21-26
Continuous description and causality, 26-29
The intricacy of the continuum, 29-39
The makeshift of wave mechanics, 39-47
The alleged break-down of the barrier between subject and object, 47-53
Atoms or quanta - the counter-spell of old standing, to escape the intricacy of the continuum, 53-58
Would physical indeterminacy give free will a chance? 58-64
The bar to prediction, according to Niels Bohr, 64-67
Literature, 68

9.2 *Naturwissenschaft und Humanismus*. Die heutige Physik
[Translation into German by E. Schrödinger]
Wien: Deuticke. 1951.

9.3 *Scienza e umanesimo*. La fisica del nostro tempo
[Translation into Italian by P. Lantermo]
Firenze: Sansoni. 1953.

9.4 *Scienza e umanesimo*. La fisica del nostro tempo
[with *Che cos'è la vita?* [B 5.8], translated into Italian by P. Lantermo and M. Ageno, added]
Firenze: Sansoni. 1970 and 1978.

9.5 *Ciencia y humanismo*. La física en nuestro tiempo
[Translation into Spanish by I. Bolívar]
Madrid: Alhambra. 1954.

9.5a *Ciencia y humanismo*
[Translation into Spanish by F. Martín]
Barcelona: Tusquets. 1985. (Serie metatemas 10, Cuadernos infimos 126)

9.6 *Science et humanisme*. La physique de notre temps
[Translation into French by J. Ladrière]
Bruges: Desclée de Brouwer. 1954.
Paris: Desclée de Brouwer. 1954.

9.7 *Kagaku to Hyûmanizumu*
[Translation into Japanese by Kôji Fushimi et al.]
Tôkyô: Misuzu shobô. 1956.

9.8 *Naturvidenskab og humanisme*
[with *Bevidsthed og materie* [B 14.5] translated into Danish by C. H. Koch, added]
København: Munksgaard. 1965.

1954

10.1 *Nature and the Greeks*. Shearman Lectures, delivered at University College, London on 24, 26, 28, and 31 May 1948
Cambridge: University Press. 1954.
Toronto: Macmillan. 1954.
The motives for returning to ancient thought, 1-19 [B 12]
The competition, reason v. senses, 20-31
The Pythagoreans, 32-50
The Ionian enlightenment, 51-66
The religion of Xenophanes. Heraclitus of Ephesus, 67-72
The atomists. 73-87 [A 196]
What are the special features? 88-96 [B 12]

10.2 *Die Natur und die Griechen*. Kosmos und Physik
[Translation into German by M. Koffka]
Wien: Zsolnay. 1955.
Hamburg: Rowohlt. 1956. (rowohlts deutsche enzyklopädie 28)
Hamburg, Wien: Zsolnay. 1959. (Jubiläumsausgabe)
Reprint: Wien, Hamburg: Zsolnay. 1983. 1987.
Zürich: Diogenes. 1989. (detebe 21781)

10.3 *La naturaleza y los griegos*
[Translation into Spanish by F. Portillo]
Madrid: Aguilar. 1961.

10.4 *Al-Tabi' ah wa-al-Ighriq*
[Translation into Arabic by 'Izzat Qurani]
al-Qahirah: Dar al-Nahdah al-'Arabiyah. 1963.

1956

11 *Expanding Universes*
Cambridge: University Press. 1956.
Preface
The de Sitter Universe, 1-40
The Theory of Geodesics, 41-64
Waves in General Riemannian Space-time, 65-74
Waves in an Expanding Universe, 75-92
Bibliography, 93

12 *What is Life? and Other Scientific Essays*
New York: Doubleday. 1956. (Doubleday Anchor Book A 88)
What is Life? 1-88 [B 5.1]
Nature and the Greeks, 89-109 [reprinted are the first and the seventh chapter of B 10.1]
Science and Humanism, 110-117 [reprinted are the first two chapters of B 9.1]
The Future of Understanding, 118-131 [A 167]
Are There Quantum Jumps? 132-160 [A 176, A 177]
Our Conception of Matter, 161-177 [A 179]
On the Peculiarity of the Scientific World-View, 178-228 [A 161]
The Spirit of Science, 229-250 [A 159]

1957

13.1 *Science Theory and Man*
[Unmodified reprint of the book *Science and the Human Temperament* [B 4.1, B 4.2] published in 1935, with the essay *What is an Elementary Particle?* [A 165] added]
New York: Dover Publications. 1957.
London: Allen and Unwin. 1958.

13.2 *Elm, Nazariye va Ensan*
[Translation into Persian by Ahmade Aram]
Teheran: Enteshar. 1970.

1958

14.1 *Mind and Matter*. The Tarner Lectures, delivered at Trinity College, Cambridge, in October 1956
Cambridge: University Press. 1958. [B 18a.1]
Reprint: Cambridge: University Press. 1959. 1967.
The Physical Basis of Consciousness, 1-15, [A 198]
The Future of Understanding, 16-35 [A 167]
The Principle of Objectivation, 36-51
The Arithmetical Paradox. The Oneness of Mind, 52-68
Science and Religion, 69-87
The Mystery of the Sensual Qualities, 88-104

14.2 *La mente y la materia*
[Translation into Spanish by F. F. Santos]
Madrid: Taurus. 1958.

14.2a *Mente y Materia*
[Translation into Spanish by J. Wagensberg]
Barcelona: Tusquets, 1983. 1984. (Serie metatemas 2, Cuadernos infimos 110)

14.3 *Geist und Materie*
[Translation into German by W. Westphal]
Braunschweig: Vieweg. 1959. (Die Wissenschaft 113)
Braunschweig: Vieweg. 1961 and 1965.
Wien, Hamburg: Zsolnay. 1986.
Zürich: Diogenes. 1989. (Detebe 21782)

14.4 *Ånd og materie sett fra en fysikers synspunkt*
[Translation into Norwegian by E. Alnæs]
Oslo: Dreyer. 1965. (Perspektivbokene. Aktuell viten 8)

14.5 *Bevidsthed og materie*
[with *Naturvidenskab og humanisme* [B 9.8], translated into Danish by C. H. Koch, added]
København: Munksgaard. 1965.

14.6 *Seishin to busshitsu.* Ishiki to kagakuteki sekaizô o meguru kôsatsu
[Translation into Japanese by R. Nakamura]
Tôkyô: Kôsakusha. 1987.

14.7 *L'esprit et la matière*
[Translation into French, notes and elision by M. Bitbol]
Paris: Éditions du Seuil. 1990.

1961

15.1 *Meine Weltansicht*
Hamburg, Wien: Zsolnay. 1961. (Jubiläumsausgabe) [B 22.1]
Frankfurt: Fischer. 1963. (Fischer Bücherei des Wissens 562)
Vorwort, 7-9
Suche nach dem Weg (Vom Herbst 1925)
Über Metaphysik im Allgemeinen, 13-19
Eine unerfreuliche Bilanz, 20-24
Das philosophische Staunen, 25-28
Das Problem Ich-Welt-Tod-Vielheit, 29-38
Die vedântische Grundansicht, 39-45
Exoterische Einführung in das naturwissenschaftliche Denken, 46-56
Weiteres über die Nicht-Vielheit, 57-68
Bewußtsein, Organisch, Anorganisch, Mneme, 69-79
Über das Bewußtwerden, 80-90 [A 198]
Über das Sittengesetz, 91-101 [A 198]
Was ist wirklich (Von 1960)
Gründe für das Aufgeben des Dualismus, 105-114 [A 207h]
Innewerden der Weltgemeinschaft durch die Sprache, 115-136
Unvollkommenheit der Verständigung, 137-152
Die Identitätslehre: Licht und Schatten, 153-170
Die zwei Anlässe zum Staunen. Ersatzethik. 171-179

15.2 *My view of the world*
[Translation into English by C. Hastings]
Cambridge: University Press. 1964.
Woodbridge: OxBow. 1983.

15.3 *Ma conception du monde*
[Translation into French by C. Renova and B. Chabot]
Le Mail: Mercure de France. 1982. (Science et Conscience)

1962

16.1 *Was ist ein Naturgesetz?* Beiträge zum naturwissenschaftlichen Weltbild
München, Wien: Oldenbourg. 1962.
Reprint: München, Wien: Oldenbourg. 1967. 1979. 1987. (scientia nova)
Was ist ein Naturgesetz? 9-17 [A 69]
Die Wandlung des physikalischen Weltbegriffs, 18-26 [A 206]
Die Besonderheit des Weltbilds der Naturwissenschaft, 27-85 [A 161]
Der Grundgedanke der Wellenmechanik, 86-101 [A 101]
Unsere Vorstellung von der Materie, 102-120 [A 179]
Was ist ein Elementarteilchen? 121-143 [A 165]

16.2 *¿Qué es una ley de la naturaleza?*
[Translation into Spanish by J. J. Utrilla]
México: Fondo de Cultura Económica. 1972.

1963

17.1 *Schrödinger - Planck, Einstein, Lorentz. Briefe zur Wellenmechanik*
[Published on behalf of the Austrian Academy of Sciences by K. Przibram]
Wien: Springer. 1963.

17.2 *Letters on wave mechanics: Einstein, Schrödinger, Planck, Lorentz*
[Translation into English and introduction by M. J. Klein]
New York: Philosophical Library. 1967.

17.3 *Hadô rikigaku keiseishi. Shurêdingâ no shokan to shôden*
[Translation into Japanese together with comments, biography and bibliography by H. Ezawa]
Tôkyô: Misuzu shobô. 1982.

18 *L'immagine del mondo*
[Schrödinger himself selected the papers for this book. Translation into Italian by A. Verson]
Torino: Boringhieri. 1963. (Grafica moderna)

1967

18a.1 *What is life? Mind and Matter*
[Combined reprint of B 5.1 and B 14.1]
Cambridge: University Press. 1967.
Reprint: Cambridge: University Press. 1969. 1974. 1977. 1979. 1980. 1983.

18a.2 *Šta je život? Um i materija*
[Translation into Serbo-Croatian by P. Grujic]
Beograd: VUK Karadžic. 1980. (Biblioteka Zodijak 47)

18a.3 *Ce este viata? Spirit si materie*
[Translation into Romanian by V. Efimov, introduction by A. Glodeanu]
Bucuresti: Editura politicá. 1980. (Idei contemporane)

1970

19 *Válogatott tanulmányok*
[Selected writings, translated into Hungarian by I. Nagy]
Budapest: Gandolat Kiado. 1970.

1971

20 *Novye puti v fizike*
[Selected writings, translated into Russian by A. G. Baronov e.a.]
Moskva: Nauka. 1971.

1982

21 *Collected Papers on Wave Mechanics. Four Lectures on Wave Mechanics*
[Third edition of B 1.6 together with B 2.1]
New York: Chelsea Publishing Company. 1982.

1985

22.1 *Mein Leben, Meine Weltansicht*
[First publication of Schrödinger's autobiography together with a reprint of *Meine Weltansicht* [B 15.1] and an introduction by A. Dick]
Wien, Hamburg: Zsolnay. 1985.
Zürich: Diogenes. 1989. (Detebe 21783)

22.2 *Shurêdingâ waga sekaikan (jiden)*
[Translation into Japanese by R. Nakamura]
Tôkyô: Kyôritsu Shuppan. 1987.

22.3 *La mia visione del mondo. La mia vita. Poesie*
[Translation into Italian and presentation by B. Bertotti. All poems of B 7.1 were reprinted and the larger part of them translated by either H. Fachinelli or E. V. de Regny]
Milano: Garzanti. 1987. (Saggi rossi)

22.4 *Mi concepcion del mundo. Mi vida.*
[Translation into Spanish by J. Fingerhut and A. Klein]
Barcelona: Tusquets. 1988. (Serie metatemas 16, Superinfimos 12)

Auguste Dick
Gabriele Kerber
Wolfgang Kerber
Karl von Meyenn

NAME INDEX

SUBJECT INDEX

THEORY AND DECISION LIBRARY

SERIES A: PHILOSOPHY AND METHODOLOGY OF THE SOCIAL SCIENCES

Editors: W. Leinfellner (*Vienna*) and G. Eberlein (*Munich*)

1. G. Zecha and P. Weingartner (eds.): *Conscience: An Interdisciplinary View*. Salzburg Colloquium on Ethics in the Sciences and Humanities (1984). 1987 ISBN 90-277-2452-0
2. R.W. Scholz: *Cognitive Strategies in Stochastic Thinking*. 1987 ISBN 90-277-2454-7
3. H. Nurmi: *Comparing Voting Systems*. 1987 ISBN 90-277-2600-0
4. M. Schmid and F.M. Wuketits (eds.): *Evolutionary Theory in Social Science*. 1987 ISBN 90-277-2612-4
5. C.L. Sheng: *A New Approach to Utilitarianism*. A Unified Utilitarian Theory and Its Application to Distributive Justice. 1991 ISBN 0-7923-1301-1
6. F. Forman: *The Metaphysics of Liberty*. 1989 ISBN 0-7923-0080-7
7. G. Bernard: *Principia Economica*. 1989 ISBN 0-7923-0186-2
8. J.W. Sutherland: *Towards a Strategic Management and Decision Technology*. Modern Approaches to Organizational Planning and Positioning. 1989 ISBN 0-7923-0245-1
9. C. Vlek and G. Cvetkovich (eds.): *Social Decision Methodology for Technological Projects*. 1989 ISBN 0-7923-0371-7
10. P. Hoyningen-Huene and F.M. Wuketits (eds.): *Reductionism and Systems Theory in the Life Sciences*. Some Problems and Perspectives. 1989 ISBN 0-7923-0375-X
11. K.G. Grunert and F. Ölander (eds.): *Understanding Economic Behaviour*. 1989 ISBN 0-7923-0482-9
12. G. Antonides: *The Lifetime of a Durable Good*. An Economic Psychological Approach. 1990 ISBN 0-7923-0574-4
13. G.M. von Furstenberg (ed.): *Acting under Uncertainty*. Multidisciplinary Conceptions. 1990 ISBN 0-7923-9063-6
14. R.L. Dukes: *Worlds Apart*. Collective Action in Simulated Agrarian and Industrial Societies. 1990 ISBN 0-7923-0620-1
15. S.O. Funtowicz and J.R. Ravetz: *Uncertainty and Quality in Science for Policy*. 1990 ISBN 0-7923-0799-2
16. J. Götschl (ed.): *Erwin Schrödinger's World View*. The Dynamics of Knowledge and Reality. 1992 ISBN 0-7923-1694-0

KLUWER ACADEMIC PUBLISHERS – DORDRECHT / BOSTON / LONDON

